AF360890

Unmanned Aerial Vehicles: Platforms, Applications, Security and Services

Unmanned Aerial Vehicles: Platforms, Applications, Security and Services

Editors

Carlos Tavares Calafate
Mauro Tropea

MDPI • Basel • Beijing • Wuhan • Barcelona • Belgrade • Manchester • Tokyo • Cluj • Tianjin

Editors
Carlos Tavares Calafate
Technical University of Valencia
Spain

Mauro Tropea
University of Calabria
Italy

Editorial Office
MDPI
St. Alban-Anlage 66
4052 Basel, Switzerland

This is a reprint of articles from the Special Issue published online in the open access journal *Electronics* (ISSN 2079-9292) (available at: https://www.mdpi.com/journal/electronics/special_issues/UAV_Platform_Applications).

For citation purposes, cite each article independently as indicated on the article page online and as indicated below:

LastName, A.A.; LastName, B.B.; LastName, C.C. Article Title. *Journal Name* **Year**, *Article Number*, Page Range.

ISBN 978-3-03936-708-5 (Hbk)
ISBN 978-3-03936-709-2 (PDF)

Contents

About the Editors

Carlos Tavares Calafate (Full Professor) is a member of the Department of Computer Engineering at the Technical University of Valencia (UPV) in Spain. He graduated with honors in Electrical and Computer Engineering at the University of Oporto (Portugal) in 2001. He completed his PhD in Informatics at the Technical University of Valencia in 2006, where he has worked since 2002. His research interests include ad hoc and vehicular networks, UAVs, smart cities and IoT, QoS, network protocols, video streaming, and network security. To date, he has published more than 400 articles, several of which have been journals, including IEEE Transactions on Vehicular Technology, IEEE Transactions on Mobile Computing, IEEE/ACM Transactions on Networking, Elsevier Ad Hoc Networks and IEEE Communications Magazine. He is an associate editor for several international journals, and has participated in the TPC of more than 250 international conferences. He is ranked among the top 100 Spanish researchers in the Computer Science and Electronics field. He is also a founding member of the IEEE SIG on Big Data with Computational Intelligence.

Mauro Tropea (Post Doc Researcher) works in the DIMES Department at the University of Calabria in Italy. He received his master's degree in April 2003 and PhD in January 2009, both in computer science engineering, at the University of Calabria, Cosenza, Italy. From May 2008 to October 2008, he was a visiting researcher at the Telecommunication Department of ESA ESTEC Noordwijk, The Netherlands. Since 2003, he has been with the telecommunications group of the DIMES Department, and, since September 2018, he has been a postdoctoral researcher there. His research interests include satellite communications, QoS architectures, bio-inspired algorithms, VANETs, FANETs, hierarchical networks, and multicasting.

Preface to "Unmanned Aerial Vehicles: Platforms, Applications, Security and Services"

In the years to come, UAVs are expected to keep gaining momentum and be adopted for an ever-growing number of tasks in different fields. This creates multiple challenges in terms of system/navigation, requiring significant integration efforts, multiple testbeds and deployment results, and novel protocols. In the enabling of such systems, communications play a vital role, and so, issues like software-defined radio/networks, virtualized networks, heterogeneous networks and channel modeling will be key to make these systems possible, especially if we keep in mind the trend towards more network demanding applications, like video streaming for real-time monitoring. Moreover, issues like UAV identification, authentication and network security always remain critical factors, especially when the UAVs are deployed to provide aerial surveillance or civil security, among other things.

In this book, we present a collection of contributions from different authors that represent an advancement in different areas related to UAVs. In particular, the first chapter, entitled "Design and Analysis of Refined Inspection of Field Conditions of Oilfield Pumping Wells Based on Rotorcraft UAV Technology", deals with an oil well monitoring method. The authors propose the use of computer vision in the detection of working conditions in oil extraction, by making use of UAV aerial photography images combined with the YOLOv3 framework for tracking detection. Through different experiments, they prove the benefits of their proposal. The second chapter, entitled "Accurate Landing of Unmanned Aerial Vehicles Using Ground Pattern Recognition", presents a solution for high precision landing. The authors propose a vision-based landing solution that relies on ArUco markers that allow the UAV to detect the exact landing position from a high altitude (30 m). They evaluate their system through a platform based on Arduino hardware, and they show how their proposal improves the landing accuracy (offset of about 11 cm) compared to the traditional GPS-based one, whose offset is about 1–3 meters. The third chapter, entitled "An Efficient and Provably Secure Certificateless Blind Signature Scheme for Flying Ad-Hoc Network Based on Multi-Access Edge Computing", proposes an efficient and provably secure certificateless blind signature scheme (CL-BS), based on multi-access edge computing (MEC) for a FANET environment, using the concept of hyperelliptic curve. The scope of the paper involves the resolution of computational and communication issues of the existing security approaches. The authors propose the use of multi-access edge computing (MEC) in a UAV environment, with the help of the 5G mobile network enabling a secure communication between UAVs and the base station (BS). The fourth chapter, entitled "A Traceable and Privacy-Preserving Authentication for UAV Communication Control System", proposes a traceable and privacy-preserving authentication to integrate the elliptic curve cryptography (ECC), digital signature, hash function, and other cryptography mechanisms for UAV application. The authors designed a traceable and privacy protection protocol to conduct the UAVs' application in a sensitive control area. This study also analyzed the computation and communication costs, to prove that the proposed scheme is practical in the real world. The fifth chapter, entitled "A New FANET Simulator for Managing Drone Networks and Providing Dynamic Connectivity", deals with the possibility of providing wireless connectivity, using a flying ad hoc network (FANET) in all those emergency situations where the traditional network can encounter several difficulties. A software simulator is proposed to implement different models: footprint, human mobility and drone behavior. The sixth chapter, entitled "Onboard Visual Horizon Detection for Unmanned Aerial

Systems with Programmable Logic", introduces a fast horizon detection algorithm suited for visual applications, to be used on board a small unmanned aircraft. For this purpose, the designed algorithm has a low complexity, in order to meet the power consumption requirements and to keep the computational cost low. The authors present formulae for distorted horizon lines. The performance of the proposed algorithm is tested on a real flight with the help of a FPGA implementation. Finally, the seventh chapter, entitled "LoRaWAN Networking in Mobile Scenarios Using a WiFi Mesh of UAV Gateways", proposes a double-layer network system called LoRaUAV. The system is based on an ad hoc WiFi network of unmanned aerial vehicle (UAV) gateways able to act as relay for the traffic generated between mobile LoRaWAN nodes and a remote base station (BS). The core of the system is a completely distributed mobility algorithm, based on virtual spring forces, that periodically updates the UAV topology to adapt to the movement of ground nodes. The proposed system is implemented in NS-3, and the performance, evaluated in a wild area firefighting scenario, shows the improvement in terms of the packet reception ratio (PRR).

Carlos Tavares Calafate, Mauro Tropea
Editors

Editorial

Unmanned Aerial Vehicles—Platforms, Applications, Security and Services

Carlos T. Calafate [1],* and Mauro Tropea [2]

[1]   Department of Computer Engineering (DISCA), Universitat Politècnica de València, 46022 València, Spain
[2]   DIMES Department, University of Calabria, 87036 Rende (CS), Italy; m.tropea@dimes.unical.it
*   calafate@disca.upv.es

Received: 9 June 2020; Accepted: 9 June 2020; Published: 11 June 2020

1. Introduction

The use of unmanned aerial vehicles (UAVs) has attracted prominent attention from researchers, engineers, and investors in multidisciplinary fields such as agriculture, signal coverage, emergency situations, disaster events, farmland and environment monitoring, 3D-mapping, and so forth. The use of this technology is playing an important role in supporting human activities. Man is concentrating more and more on intellectual work, trying to automate practical activities as much as possible in order to increase their efficiency. In this regard, the use of drones is increasingly becoming a key aspect of this automation process. A drone offers many advantages including agility, efficiency and reduced risk, especially in dangerous missions. Hence, this special issue focuses on applications, platforms and services where UAVs can be used as facilitators for the task at hand, also keeping in mind that security should be addressed from its different perspectives, ranking from communications security to operational security, and also keeping in mind privacy issues.

2. The Present Issue

In response to the call for papers, we received 11 submissions, and 7 of these manuscripts have been accepted for publication.

The first paper, titled "LoRaWAN Networking in Mobile Scenarios Using a WiFi Mesh of UAV Gateways" [1], proposes a double-layer network system called LoRaUAV. The system is based on a WiFi ad hoc network of Unmanned Aerial Vehicle (UAV) gateways able to act as relay for the traffic generated between mobile LoRaWAN nodes and a remote Base Station (BS). The core of the system is a completely distributed mobility algorithm based on virtual spring forces that periodically updates the UAV topology to adapt to the movement of ground nodes. The proposed system is implemented in NS-3 and the performance, evaluated in a wild area firefighting scenario, shows the improvement in terms of Packet Reception Ratio (PRR).

The second paper, entitled "Onboard Visual Horizon Detection for Unmanned Aerial Systems with Programmable Logic" [2], introduces a fast horizon detection algorithm suited for visual applications to be used on board a small unmanned aircraft. For this purpose, the designed algorithm has a low complexity in order to meet the power consumption requirements and to keep the computational cost low. The authors present formulae for distorted horizon lines. The performance of the proposed algorithm is tested on a real flight with the help of a FPGA implementation.

The third paper, entitled "A New FANET Simulator for Managing Drone Networks and Providing Dynamic Connectivity" [3], deals with the possibility of providing wireless connectivity using a flying ad-hoc network (FANET) in all those emergency situations where the traditional network can meet several difficulties. A software simulator is proposed implementing different models—footprint, human mobility and drone behavior.

The fourth paper, entitled "A Traceable and Privacy-Preserving Authentication for UAV Communication Control System" [4], proposes a traceable and privacy-preserving authentication to integrate the elliptic curve cryptography (ECC), digital signature, hash function, and other cryptography mechanisms for UAV application. The authors designed a traceable and privacy protection protocol to conduct the UAVs' application in a sensitive control area. This study also analyzed the computation and communication cost to prove the proposed scheme is practical in the real world.

The fifth paper, entitled "An Efficient and Provably Secure Certificateless Blind Signature Scheme for Flying Ad-Hoc Network Based on Multi-Access Edge Computing" [5], proposes an efficient and provably secure certificateless blind signature scheme (CL-BS) based on multi-access edge computing (MEC) for a FANET environment using the concept of hyperelliptic curve. The scope of the paper is to resolve computational and communication issues of the existing security approaches. The authors propose the use of multi-access edge computing (MEC) in a UAV environment with the help of 5G mobile network enabling a secure communication between UAVs and the base station (BS).

The sixth paper, entitled "Accurate Landing of Unmanned Aerial Vehicles Using Ground Pattern Recognition" [6], presents a solution for high precision landing. The authors propose a vision-based landing solution that relies on ArUco markers that allow the UAV to detect the exact landing position from a high altitude (30 m). They evaluate their system through a platform based on Arduino hardware, and they show how their proposal improves the landing accuracy (offset of about 11 cm) compared to the traditional GPS-based one, whose offset is about 1–3 m.

Finally, the seventh and last paper, entitled "Design and Analysis of Refined Inspection of Field Conditions of Oilfield Pumping Wells Based on Rotorcraft UAV Technology" [7], deals with an oil well monitoring method. The authors propose using computer vision in the detection of working conditions in oil extraction by making use of UAV aerial photography images combined with the YOLO v3 framework for tracking detection. Through different experiments they prove the goodness of their proposal.

3. Future

In the next years UAVs are expected to keep gaining momentum, being adopted for an ever-growing number of tasks in different fields. This entangles multiple challenges in terms of system/navigation, requiring significant integration efforts, multiple testbeds and deployment results, and novel protocols. To enable such systems, communications play a vital role, and so issues like software-defined radio/networks, virtualized networks, heterogeneous networks and channel modeling will be key to make these systems possible, especially if we keep in mind the trend towards more network demanding applications, like video streaming for real-time monitoring. Finally, issues like UAV identification, authentication and network security always remain as critical factors, especially when the UAVs are deployed to provide aerial surveillance or civil security, among others.

Acknowledgments: We thank all the authors for submitting their work to this Special Issue. We also thank all the reviewers, who contributed to improve the quality of the papers through their valuable comments and suggestions. We are extremely grateful to Juan-Carlos Cano, Section Editor-in-Chief of MDPI *Electronics*, for giving us the possibility of serving the community with this Special Issue, and to Michelle Zhou, the managing Editor of this Special Issue.

Conflicts of Interest: The authors declare no conflict of interest.

Electronics **2020**, *9*, 975

References

1. Stellin, M.; Sabino, S.; Grilo, A. LoRaWAN Networking in Mobile Scenarios Using a WiFi Mesh of UAV Gateways. *Electronics* **2020**, *9*, 630, doi:10.3390/electronics9040630.
2. Hiba, A.; Sántha, L.M.; Zsedrovits, T.; Hajder, L.; Zarandy, A. Onboard Visual Horizon Detection for Unmanned Aerial Systems with Programmable Logic. *Electronics* **2020**, *9*, 614, doi:10.3390/electronics9040614.
3. Tropea, M.; Fazio, P.; De Rango, F.; Cordeschi, N. A New FANET Simulator for Managing Drone Networks and Providing Dynamic Connectivity. *Electronics* **2020**, *9*, 543, doi:10.3390/electronics9040543.
4. Chen, C.L.; Deng, Y.Y.; Weng, W.; Chen, C.H.; Chiu, Y.J.; Wu, C.M. A Traceable and Privacy-Preserving Authentication for UAV Communication Control System. *Electronics* **2020**, *9*, 62, doi:10.3390/electronics9010062.
5. Khan, M.A.; Qureshi, I.M.; Ullah, I.; Khan, S.; Khanzada, F.; Noor, F. An Efficient and Provably Secure Certificateless Blind Signature Scheme for Flying Ad-Hoc Network Based on Multi-Access Edge Computing. *Electronics* **2020**, *9*, 30, doi:10.3390/electronics9010030.
6. Wubben, J.; Fabra, F.; Calafate, C.T.; Krzeszowski, T.; Marquez-Barja, J.M.; Cano, J.C.; Manzoni, P. Accurate Landing of Unmanned Aerial Vehicles Using Ground Pattern Recognition. *Electronics* **2019**, *8*, 1532, doi:10.3390/electronics8121532.
7. Zhou, Y.; Wu, C.; Wu, Q.; Eli, Z.M.; Xiong, N.; Zhang, S. Design and Analysis of Refined Inspection of Field Conditions of Oilfield Pumping Wells Based on Rotorcraft UAV Technology. *Electronics* **2019**, *8*, 1504, doi:10.3390/electronics8121504.

Article

Design and Analysis of Refined Inspection of Field Conditions of Oilfield Pumping Wells Based on Rotorcraft UAV Technology

Yu Zhou [1], Chunxue Wu [1], Qunhui Wu [2], Zelda Makati Eli [1], Naixue Xiong [1] and Sheng Zhang [1,*]

[1] School of Optical-Electrical and Computer Engineering, University of Shanghai for Science and Technology, Shanghai 200093, China; zhouyu0509@126.com (Y.Z.); wcx@usst.edu.cn (C.W.); m18097909971@163.com (Z.M.E.); xiongnaixue@gmail.com (N.X.)

[2] Shanghai HEST Co. Ltd., Shanghai 201610, China; shhest@aliyun.com

[*] Correspondence: zhangsheng_usst@aliyun.com; Tel.: +86-1338-6002-013

Received: 13 October 2019; Accepted: 28 November 2019; Published: 9 December 2019

Abstract: The traditional oil well monitoring method relies on manual acquisition and various high-precision sensors. Using the indicator diagram to judge the working condition of the well is not only difficult to establish but also consumes huge manpower and financial resources. This paper proposes the use of computer vision in the detection of working conditions in oil extraction. Combined with the advantages of an unmanned aerial vehicle (UAV), UAV aerial photography images are used to realize real-time detection of on-site working conditions by real-time tracking of the working status of the head working and other related parts of the pumping unit. Considering the real-time performance of working condition detection, this paper proposes a framework that combines You only look once version 3 (YOLOv3) and a sort algorithm to complete multi-target tracking in the form of tracking by detection. The quality of the target detection in the framework is the key factor affecting the tracking effect. The experimental results show that a good detector makes the tracking speed achieve the real-time effect and provides help for the real-time detection of the working condition, which has a strong practical application.

Keywords: computer vision; oil well working condition; real-time detection; sort; unmanned aerial vehicle (UAV); YOLOv3

---

1. Introduction

The fault diagnosis technology and working condition monitoring technology of the pumping unit have always been the focus of the oilfield. At present, the commonly used fault diagnosis methods are mainly manual analysis and indicator diagram diagnosis. However, the dependence of a large number of high-precision sensors and high-sensitive devices not only increases the original cost of working condition detection but also gradually increases the requirements of staff [1]. The whole process takes a lot of time, and even real-time working conditions cannot be obtained. This has posed a great challenge to the detection of field working conditions of oil field pumping wells [2]. In recent years, with the gradual maturity of UAV technology, more and more projects have been launched around UAV, and it has been widely used in the inspection of power, highway, agriculture, communication, oil, and other fields [3]. By making use of the flexible mobility and powerful timeliness of the UAV, the difficulty of traditional condition detection can be overcome by using the UAV patrol mode [4,5].

The subject of this paper is the fine inspection research of pumping-well working conditions based on UAV. Unmanned UAVs equipped with high-definition cameras can hover in the air for a long time to monitor the ground over a wide range and obtain real-time images. Therefore, through the pumping unit's real-time images acquired by the UAV, the deep learning detection [6,7] and the tracking method are used to detect the working condition of the oil-well pumping unit in operation. The specific detection precision is to the extent of the pumping unit's key parts [8]. At the same time of the whole pumping unit detection, the head working part of the pumping unit also undergo detailed detection and tracking, so as to achieve more refined inspection and get a more detailed pumping condition. Tracking the working state of the pumping unit and key components provides real-time position and movement information of the specified target [9]. By analyzing the state of the pumping unit and the real-time working state of the key components, the purpose of the drone's refined detection of the oil-well pumping unit is achieved [10].

Because there are multiple targets on the oil field, such as vehicles and workers, the purpose of this paper is to track multiple specified targets in the UAV image, which becomes a problem of multi-target tracking [11]. Multi-target tracking lacks artificial markers, and there are multiple targets, so it is necessary to use a target detector to detect the position of the target in the image at each moment [12]. Therefore, this paper adopts the tracking method based on detection and matching. Firstly, the detector is used to detect the static image of the oil-well pumping unit and the important parts, such as head working. Then, the static problem is extended to the dynamic problem, and the detection results of the two frames before and after are matched one by one to realize the tracking of the key working parts of the oil-well pumping unit and the pumping unit.

In this article, the main contributions are as follows. 1) A multi-target tracking framework (YLTS) for real-time tracking is proposed. It uses YOLOv3 as the detector and the sort algorithm as the tracker. In this paper, different algorithms are used as detectors to make multi-target tracking experiments for oilfield pumping units and their related components, and their accuracy and real-time tracking effects are compared. It is concluded that the use of YOLOv3 as the detector in this framework is most suitable; 2) different from the traditional method of detecting pumping unit working conditions with indicator diagrams, this paper applies the fine inspection project of UAV to the study of pumping unit working condition detection in oil production. By detecting and tracking the pumping unit and the head working part in the oil field, the position and movement information of the key components such as the head working are obtained, which provides a reliable basis for the next semantic analysis and the judgment of the working condition.so as to obtain the real-time working condition of the oil-well pumping unit.

The rest of this article is arranged as follows. Section 2 briefly reviews the research status of pumping unit working condition detection and the application status of UAV inspection. Then the related work of the model is introduced. The proposed method is described in Section 3, and experimental results and comparisons are explained in detail in Section 4. Finally, we summarize the paper and illustrate the future work in Section 5.

2. Related Works

The pumping unit has many major components, and the common faults are also complicated. In order to meet different fault inspections, the current pumping inspection methods generally involve manual collection. High-precision sensors and high-sensitivity devices are used to detect the load and displacement, current, voltage, stroke, and stroke parameters of the pumping unit. Then display the parameter values and the indicator diagram on the LCD screen. Although this method basically satisfies the basic needs of oilfields for pumping-unit monitoring, as the scale of oilfield mining is getting larger and larger, the establishment of this system is more and more difficult and expensive.

In recent years, UAVs have been widely used in the field of inspection. However, so far, the more mature inspection application of UAVs only stays in the inspection of pipelines and routes, such as highways, high-voltage power lines, and oil and natural gas pipelines. The UAV flies along the pipeline to be inspected. In the automatic flight mode, the built-in high-definition camera is used to point at the pipeline to be inspected to collect the image of pipeline details, which is then transmitted to the ground station through wireless remote real-time transmission. In this paper, the application of UAV inspection is extended to the fine inspection of the working condition of the oil field pumping unit, so as to obtain the position of the pumping unit and the motion information of key parts in the video sequence in the middle and low altitude flight, providing a basis for further semantic layer analysis (motion state recognition, scene recognition, etc.) [13]. In this way, the real-time working condition of the oil-well pumping unit can be further judged according to the obtained information.

In order to achieve the work status tracking for pumping units key component, based on the requirement of real-time and multi-target, the technology adopted in this paper is the target tracking algorithm based on detection and matching. The detection quality in this method largely affects the tracking effect, so the key technology of this algorithm lies in the image target detection algorithm of deep learning. This chapter mainly introduces the main algorithms and related concepts used in this paper, including the principle of convolutional neural networks (CNN) in deep learning and the most advanced algorithms in the field of image detection, and time series prediction algorithms.

2.1. The Basics of Convolutional Neural Networks (CNNs)

The convolutional neural network (CNN) is a deep learning algorithm, which is an application of deep learning algorithms in the field of image processing and has excellent performance for large-scale image processing [14]. Inspired by the biological neural network, the perception layer was used to simulate the process of obtaining image information in biological vision, the hidden layer was used to simulate the neurons in the biological neural network, and the convolutional layer and excitation function were used to simulate the process of information transmission between neurons in the biological neural network. CNN uses a large number of hidden nodes to store the data of the original image. This method can obtain a better representation than the original image, and the tile processing method of hidden layer nodes makes the CNN have translation invariance. The schematic diagram of a CNN is shown in Figure 1:

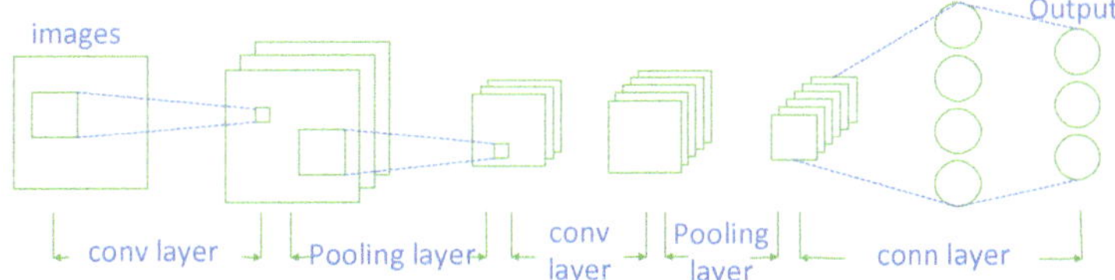

Figure 1. The basic construction of a convolutional neural network (CNN).

As shown in Figure 1, a CNN is made up of several convolution layers [15], a pooling layer, and a fully connected layer. Multiple convolutional layers are accompanied by a pooling layer. After repeated cycles, a fully connected layer is added to form a CNN. The convolution layer is the layer responsible for the transformation from the real domain to the feature domain, and it is also the most critical layer. The purpose of the pooling layer is to subsample the convolution result [16], extract the important part of the feature, reduce the number of network parameters, prevent the emergence of an over-fitting image, and improve the robustness of the network. The fully connected layer is mainly used to make some local features have global characteristics. All neuron nodes in this layer will be connected with the output of all neurons in the convolution layer of the previous Layer. Therefore, the calculation amount of the fully connected layer is relatively large. The output result of the fully connected layer will be taken as the input of the classifier [17].

2.2. Object Detection

Object detection refers to detecting the location of objects in an image while classifying images. The deep convolutional neural network (DCNN) has made great achievements in image object detection after face recognition. In recent years, a large number of efficient object detection algorithms based on deep learning have emerged successively, such as the region-convolutional neural network

(R-CNN), fast region-convolutional neural network (Fast R-CNN), faster region-convolutional neural network (Faster R-CNN), You only look once (YOLO), and Single Shot Multi-Box Detector (SSD) [18]. These algorithms are divided into two categories according to whether there is a region proposal.

2.2.1. Faster R-CNN

Faster R-CNN is the most advanced algorithm for object detection in R-CNN series images based on deep learning. It introduced the region proposal network (RPN) to directly generate candidate regions, which can be seen as a combination of the RPN and Fast R-CNN model [19].

For the RPN, a CNN model (commonly known as a feature extractor) is used to receive the whole picture and extract the feature graph. An N × N sliding window is then used on the feature graph to map a low-dimensional feature (e.g., 256–d) for each sliding window position. This feature is then fed into two fully connected layers, one for classification prediction and one for regression. For each window position is a set k different size or scale of a priori box (anchors, default bounding boxes), which means that each location has a prediction k candidate region (region proposals). For the classification layer, its output size is 2k, which represents the probability value that each candidate region contains object or background, while the regression layer outputs 4k coordinate values, which represents the position of each candidate region (relative to each prior box). The two full connection layers are shared for each sliding window location. Therefore, RPN can be realized by convolution layer: firstly, an n × n convolution to obtain low-dimensional features, and then two 1 × 1 convolutions for classification and regression, respectively. The network architecture of RPN is shown in Figure 2.

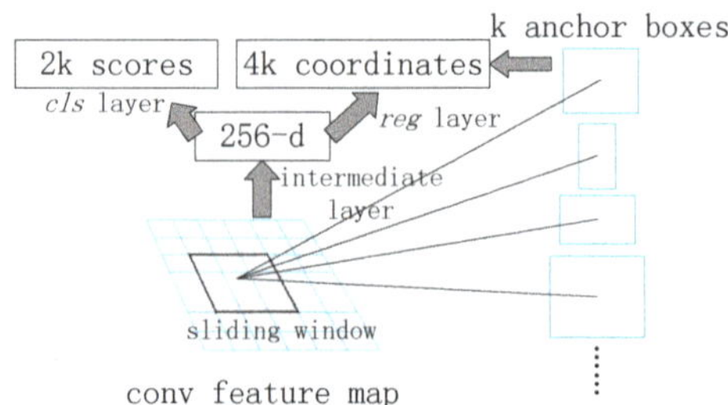

Figure 2. Region proposal network (RPN) network architecture.

The region proposal network uses dichotomies to distinguish only the background and objects but does not predict the categories of objects, namely class-agnostic. This method solves the regional recommendation and time-consuming problems in Fast R-CNN and greatly improves the detection speed. The detection of the mean average precision (mAP) value of PASCAL VOC 2007 increased from 70% to 73.2%.

2.2.2. YOLO

The YOLO neural network is based on the regression method to complete the target detection instead of the regional recommendation. It was proposed by Ross et al. in 2015 [20], which mainly transforms the multi-classification problem into a regression problem to solve the image detection. The classification and localization problems are solved by the same regression algorithm, which greatly improves the detection speed and achieves real-time effects in the field of general image target detection. YOLO first divides the whole picture into S × S grids. Each grid is responsible for predicting the position of the target point where the center point falls in this grid area. The predicted value is compared with the real value to calculate the predicted loss. The core idea is to directly operate on the entire picture, input a picture, and directly derive the position of the prediction frame and the category to which the prediction frame belongs in the output layer. Each grid into which YOLO is divided is responsible for predicting some detection frames. Each detection frame needs to have a confidence value of a specific target in addition to its own position information.

By means of direct regression of the whole graph, YOLO can greatly improve the detection speed, reduce the error rate of background prediction, and learn highly generalized features, which is better than Fast RCNN in migration learning. However, the disadvantage is that the detection accuracy is low, object positioning errors easily occur, and the detection effect on small objects is not good enough. A series of YOLO algorithms have appeared (e.g., YOLOv2, YOLOv3) in recent years and have improved and strengthened the shortcomings of the original version. Based on the research of this paper focusing on real-time and multi-objective features, the detection part used in this paper is the latest YOLOv3 neural network in this series. The use of YOLOv3 neural network algorithm modeling to implement the detector portion of this article will be described in detail in Section 3.

2.2.3. SSD

The Single Shot Multi-Box Detector (SSD) belongs to the multi-box prediction of a one-stage method. The main idea is to carry out dense sampling uniformly on the feature graph of multiple layers in the image [21]. Different scales and aspect ratios can be adopted in sampling, and then features can be extracted by CNN for classification and regression. The whole process only takes one step, so it has the advantage of fast speed. However, an important disadvantage of uniform dense sampling is that training is difficult, mainly because the positive sample and the negative sample (background) are extremely unbalanced, resulting in slightly lower accuracy of the model [22].

Given the advantages and disadvantages of the RCNN series and the YOLO series, the SSD algorithm borrows many of these ideas and has many ideological improvements. Respectively, they are:

1. Multi-scale feature graph is adopted for detection—pyramid feature.
2. Set Default boxes.
3. Determination of Default boxes size.
4. Convolution was used for detection.

The above improvements made the detection speed faster than YOLOv1 and the accuracy faster than Faster R-CNN. However, the initial size and aspect ratio of the default boxes need to be set manually, and the size and shape of the default box used by the feature of each layer in the network are just different, which makes the debugging process very dependent on experience [23]. Moreover, the recognition of small-size objects is still poor, which cannot reach the level of Faster R-CNN. In contrast, the YOLOv3 used in this paper has obvious advantages in small object detection after absorbing the advantages and disadvantages of the first two versions and is much faster than SSD. This is one of the reasons why this article uses YOLOv3 instead of SSD as a detector.

3. Using the YLTS Framework to Realize the Pumping Unit Working Condition Detection of the Aerial Image of the UAV

This paper uses the proposed YLTS framework to achieve multi-target tracking [24,25]. Before the tracking, YOLOv3 was used to complete the detection of all the pumping units and the head working parts in the video to realize feature modeling, and then, the sorting tracking algorithm to complete the multi-target tracking was used. The whole process was to achieve multi-target tracking by detecting and then using prediction and matching. The framework proposed in this paper achieves real-time tracking, but mainly depends on the performance of the detector in the framework. YOLOv3, as a target detector, was a relatively good model in recent years. After experimental comparison, it is concluded that the use of YOLOv3 as a detector enables the framework to achieve faster real-time effects in tracking speed. Because the state of the UAV is in cruise, the main purpose of this article concerns the low altitude cruise in the detection of the oil pumping unit and the head working, and mainly discusses the work condition of the head working (work cycle, movement speed, movement direction) for real-time tracking, access to the above information can be used according to its working status for further analysis of the pumping unit working condition.

3.1. Using YOLOv3 as a Detector of the YLTS Framework to Detect the Pumping Unit and the Head Working

In order to learn more about YOLOv3, the first two versions of YOLO (v1, v2) must be understood first. Since many of YOLOv3's ideas are inherited from v1 and v2, this section first introduces YOLOv1, and then introduces YOLOv3 in detail.

The earliest version of the YOLO series is YOLOv1, which is a detection model that converts multiple classification problems into regression problems for solution. The classification and location problems in the detection of a pumping unit and head working are solved by the same regression algorithm, which greatly improves the detection speed. It uses a separate CNN model to realize end-to-end target detection, divides the input images into 7×7 grids, and then each cell is responsible for predicting the targets in which the center points fall in the grid; when the pumping unit or head working fall in some grid, this grid is responsible for predicting them, compares the predicted value with the real value, and calculates the predicted loss. The core idea is to directly manipulate the whole picture by inputting a figure directly in the output layer for each grid to predict the B bounding box location information and the confidence score of the bounding box [26].

The predicted value of each bounding box contains five elements: (x,y,w,h,c), where (x,y) represents the center coordinate of the boundary box, and the predicted value (x,y) of the center coordinate is the offset value relative to the coordinate point in the upper left corner of each cell; w and h are the width and height of the bounding box, and the predicted values of w and h of the bounding box are the ratio of the width and height relative to the entire image, and the value c is confidence score. The confidence score includes two aspects: on the one hand, the probability of the boundary box containing the target is denoted as $Pr(object)$; if the pumping unit or the head working part in the picture falls in the grid cell, it is set as 1, otherwise, it is 0. On the other hand, the accuracy of the boundary box can be represented by the intersection ratio (IOU) of the prediction box and ground truth, denoted as IOU_{pred}^{truth}, so the confidence is defined as $Pr(object) * IOU_{pred}^{truth}$. The multiplication of confidence scores and conditional probability is the solution of the classification problem, such as Formula (1):

$$Pr(class_i|object) * Pr(object) * IOU_{pred}^{truth} = Pr(class_i) * IOU_{pred}^{truth}. \tag{1}$$

As shown in Formula (1), it represents the confidence of the category. In the classification problem, each grid unit also predicts C conditional category probabilities $Pr(Class|Object)$ that are conditional on the inclusion of the target grid unit. Each grid cell predicts only one set of category probabilities, regardless of the number of bounding boxes B. This paper aims to detect the pumping unit and head working parts in the field with complicated environmental conditions, so C here is set as 2, which also reduces the workload of the algorithm. In the test, the conditional class probability is multiplied by the predicted confidence value of each box, so as to calculate the class-specific confidence scores of each boundary box, what it represents is the probability that the target belongs to a pumping unit or head working in the boundary box and the quality that the boundary box matches the target. Prediction boxes of the network are generally filtered according to category confidence. In general, each cell needs to predict $(B \times 5 + C)$ values. If the input image is divided into an $S \times S$ grid, the final predicted value is a tensor of $S \times S \times (B \times 5 + C)$ size.

The structure of the YOLO network can be seen from Figure 3 [27], which uses the convolutional network to extract features and then uses the full connection layer to obtain predicted values. It can be seen that its network has 24 convolutional layers and two fully connected layers. The fully connected layer of the last layer outputs a $7 \times 7 \times 30$ tensor; this tensor stores the location information of all the detection boxes predicted by the YOLO model and the probability values that belong to a set of specific classes with the detection boxes.

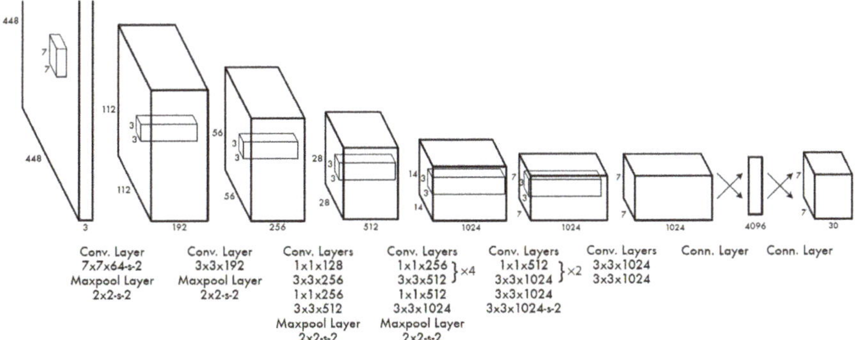

Figure 3. The YOLOv1 network structure.

The training of YOLO is end-to-end, the prediction of the position, size, type, confidence (score), and other information of the prediction box is trained by a loss function [28]. Formula (2) is YOLOv1's loss function.

$$
\begin{aligned}
loss =\ & \lambda_{coord} \sum_{i=0}^{S^2} \sum_{j=0}^{B} l_{ij}^{obj} \left[(x_i - \hat{x}_i)^2 + (y_i - \hat{y}_i)^2 \right] + \\
& \lambda_{coord} \sum_{i=0}^{S^2} \sum_{j=0}^{B} l_{ij}^{obj} \left[\left(\sqrt{\omega_i} - \sqrt{\hat{\omega}_i} \right)^2 + \left(\sqrt{h_i} - \sqrt{\hat{h}_i} \right)^2 \right] + \\
& \sum_{i=0}^{S^2} \sum_{j=0}^{B} l_{ij}^{obj} (c_i - \hat{c}_i)^2 + \\
& \lambda_{noobd} \sum_{i=0}^{S^2} \sum_{j=0}^{B} l_{ij}^{noobj} (c_i - \hat{c}_i)^2 + \\
& \sum_{i=0}^{S^2} \sum_{c \in classes} (p_i(c) - \hat{p}_i(c))^2.
\end{aligned}
\tag{2}
$$

The S^2 in Formula (2) represents the number of grids, in this case, 7×7. B is the number of prediction boxes per cell, which, in this case, is 2. The value of l_{ij}^{obj} is 0 or 1, that is, whether there is a target in the cell. The value of λ_{coord} is 5 and the value of λ_{noobd} is 0.5. Formula (2) is divided into four parts:

Part 1: The first line is the loss function for position prediction. The total square error (SSE) is used.

Part 2: The second line is the loss function for width and height. The total square error is used.

Part 3: The third and fourth rows of confidence (confidence) are also the total squared error (SSE) used as a loss function.

Part 4: The fifth line is the loss function for the class probability and also uses the total square error (SSE) as the loss function.

Finally, several loss functions are added together as a loss function of YOLOv1.

Different oilfields have different environmental conditions. In the complex environment of oilfields, the pumping unit is connected to the head working. In addition, the head working is relatively small compared with the pumping unit when the UAV is flying higher. Moreover, the up and down swing of the head working in the pumping unit may lead to the overlap with the pumping unit itself. Under such complex and harsh testing conditions, YOLOv1 cannot meet the requirements of the industrial application level. YOLOv3's improvements make it an algorithm that meets the industrial application level requirements. On the surface, the core idea of YOLOv3 is basically the same as that of YOLOv1, both of which are tested by dividing cells in a square way, but the number of partitions is different. However, its improvement makes its detection effect become an excellent detector both for accuracy and speed. For example, batch normalization has been added since v2 as a method of regularization, accelerating convergence, and avoiding overfitting, connecting the BN layer and leaky

ReLu layer to the end of each convolutional layer. The use of multilevel prediction makes up for the shortcomings of the previous version of small target detection. Multi-scale training, which allows for a trade-off between speed and accuracy, makes YOLOv3 more flexible and suitable for industrial applications. Figure 4 shows the network structure of YOLOv3.

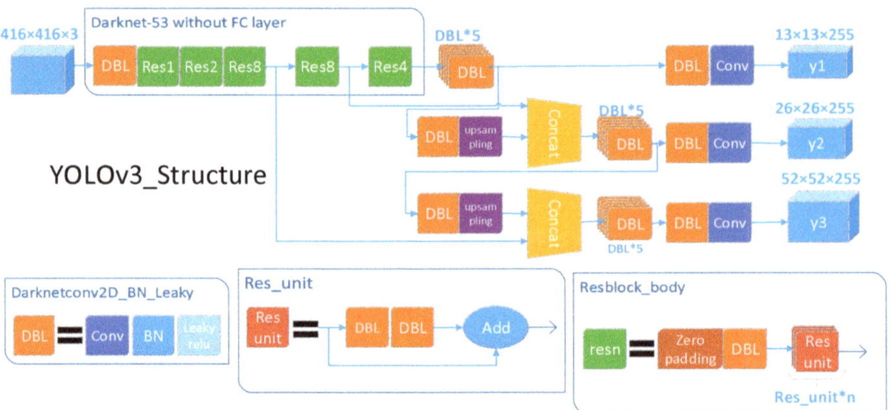

Figure 4. The YOLOv3 network structure.

Here are three additions to Figure 4:

First of all, DBL is the basic component of YOLOv3, which consists of convolution, BN, and Leaky Relu. For v3, in addition to the last layer of convolution, the three have been merged to form the smallest component. Secondly, there are multiple res, which are the big components of YOLOv3. They draw on ResNet's residual structure. Using this structure can make the network structure deeper. Its basic component is also DBL. Finally, splicing the intermediate layer of darknet and the upper sampling of a later layer. The splicing operation is different from the residual layer add operation. Splicing expands the dimension of a tensor, whereas add simply adds without changing the dimension of a tensor.

There is no pooling layer and full connection layer in the entire v3 structure, add an anchor box to predict the bounding box. This avoids the image that can only recognize the same resolution as the training image at the time of detection and can have a higher resolution at the output of the convolutional layer. It is very suitable for the occasion when the UAV is not in the fixed altitude inspection. Good detection can be maintained when the drone's flight is very close to a pumping unit or the flight altitude is high. In the process of forward propagation, the dimensional transformation of the tensor is realized by changing the step size of the convolution kernel [29]. The following analysis is carried out layer by layer.

Input layer: images are input with 416×416 pixels and 3 channels, and then the BN operation is carried out on the input. Then, the 32-layer convolution kernel operation is carried out. The size of each convolution kernel is 3×3, and the step is 1. Finally, the 416×416 feature map of 32 channels is produced as the output.

Res layer: the input and output in this layer are generally consistent, and no other operations, just subtraction. In order to solve the phenomenon of gradient diffusion or gradient explosion in deep neural network, it is proposed to change layer by layer training to stage by stage training. The deep neural network is divided into several subsegments, each of which contains a relatively shallow network layer, and then each segment is directly connected to train the residual. Each segment learns only a fraction of the total difference, and ends up with a smaller total loss. At the same time, the propagation of the gradient is well controlled to avoid situations that are not conducive to training, such as the disappearance or explosion of the gradient.

Darknet-53: from layer 0 to layer 74, there are 53 convolution layers, and the rest are res layers. This layer is the main network structure for feature extraction of YOLOv3, and the convolution layer of 3×3 and 1×1 is used [30]. A large number of jump layer connections using residuals. In the previous work, the sampling was generally conducted by max-pooling or average-pooling with the size of 2×2 and stride length of 2. However, in this network structure, convolution with a step size of 2 is used for descending sampling. At the same time, up-sampling and route operation are used in the network structure, and three times of detection are carried out in a network structure. This ensures the convergence of training. The effect of classification and detection will also be improved, and the reduction of parameters will reduce the amount of calculation. This is very good for more complex oil field sites, different locations of the sparse distribution of pumping units, plus the blocking of the head working. Better results can be obtained by using a darknet-53 network to train such complex images.

The part of YOLO: this part is divided into three scales from 75 to 105 layers, and local feature interaction is realized by a convolution kernel, such as in Figure 5.

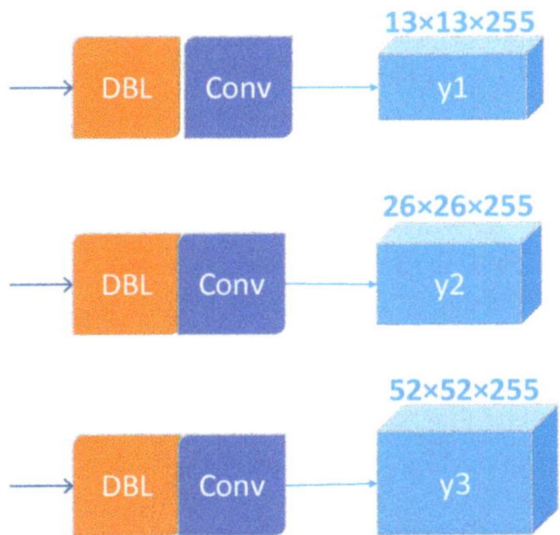

Figure 5. Output from the YOLO layer.

The minimum scale YOLO layer inputs 13×13 feature maps, a total of 1024 channels, reduces the channel to 75 by convolution operation, and finally outputs 13×13 feature maps and 75 channels, and on this basis, perform position regression and classification.

The input of the mesoscale YOLO layer is to convolve the feature map of the 13×13 and 512 channels of the 79 layer to generate the feature map of 13×13 and 256 channels. A 26×26, 256-channel feature map is generated after up sampling, and convolution is performed after merging with the 61×26, 512-channel mesoscale feature map of the 61 layer. Finally, an output of a 26×26 size feature map and 75 channels is produced.

The input of the large-scale YOLO layer is to convolve the feature map of the 91-story 26×26 and 256-channel and generate the feature map of 26×26 and 128 channels and generate the feature map of 52×52 and 128 channels after up sampling. At the same time, convolution is performed after merging with the 52×52, 256-channel mesoscale feature map of 36 layers. Finally, a feature map of size 52×52 and 75 channels are output. Based on this, position regression and classification are performed [31].

According to the structural pattern of YOLOv3, except for the last layer of the model, which uses the linear activation function, all other layers use the leaky ReLU below as the activation function:

$$y = \begin{cases} x, & x > 0 \\ 0.1x, & otherwise \end{cases} . \tag{3}$$

Compared to ordinary ReLU, leaky does not make the negative number directly 0, but multiplies it by a small coefficient (constant). Keep negative output, but reduce negative output.

Compared with YOLOv1, v3 makes some adjustments in the loss function. Except that the loss function of the width and height of the second part still uses the total square error, the loss function of other parts uses the binary cross entropy. The next step is to add them together. The loss function for V1 was explained in Formula (2) in the previous section. The following is the formula for binary cross entropy:

$$loss = -\sum_{i=1}^{n} \hat{y}_i \log y_i + (1 - \hat{y}_i), \tag{4}$$

$$\frac{\partial loss}{\partial y} = -\sum_{i=1}^{n} \frac{\hat{y}_i}{y_i} - \frac{1 - \hat{y}_i}{1 - y_i} \tag{5}$$

This is the loss function between probabilities. Only when y_i and $\hat{y}_i$ are equal, the loss will be 0; otherwise, the loss will be a positive number. Moreover, the greater the difference in probability, the greater the loss will be. This measure of probability distance is called cross entropy. YOLOv3 changes the loss function so that it can better model complex target categories and data sets of overlapping labels. It is also suitable for the data set that the head working overlaps or blocks with the pumping unit in the scene of the oil field in this paper.

Through the above modeling, the work of the detector is first completed. The detection of each frame of the pumping unit and the head working part is realized. After that, the tracker is used to complete the tracking of multiple targets.

3.2. Use the Sort Algorithm as a Tracker of the YLTS Framework to Track the Pumping Unit and the Head Working

In order to ensure the real-time tracking effect, this paper uses the Sort algorithm as a tracker to track the target based on the detector's detection of the pumping unit and the head working. The algorithm is an algorithm based on detection and multi-target tracking, which is updated online and has good real-time performance. The tracking problem is regarded as a data association problem. The Kalman filter is used to process the correlation of frame-by-frame data [32,33], and the Hungarian algorithm is used to correlate metrics. The position and size of the detection box are used to correlate the motion estimation and data of the target [34]. The following is an object state model that represents and propagates the target ID to the next frame:

$$x = \left[u, v, s, r, \dot{u}, \dot{v}, \dot{s} \right]^{T}. \tag{6}$$

where u and v represent the central coordinate of the target, s represents the size area of the target, r represents the aspect ratio of the target, which remains unchanged, and the last three quantities represent the predicted next frame.

The steps of the whole process are as follows:

1. When the first frame comes in, the detected target is initialized and a new tracker is created, labeled with an id.
2. When the following frame comes in, the state prediction and covariance prediction generated by the previous frame detection box are obtained first in the Kalman filter. The target state prediction and the IOU of the frame detection box are respectively obtained, the maximum matching of the IOU is obtained by the Hungarian assignment algorithm, and the matching pair in which the matching value is smaller than the IOU threshold is removed.
3. The Kalman tracker is updated using the matched target detection frame in this frame to calculate the Kalman gain, status update, and covariance update. The status update value is output as the tracking frame of this frame. The tracker for targets that are not matched in this frame are reinitialized [35,36].

After the above steps, the proposed YOLOv3 is used as the detector, and the sort algorithm is basically completed as the framework of the tracker. First, use YOLOv3 to test the pumping unit and the head working part, and input the test result to the tracker. As a tracker, the sort algorithm uses the Kalman filter to process the correlation of frame-by-frame data and the Hungarian algorithm to correlate metrics to track the pumping unit and the head working. After that, through the analysis of the results of the tracking, the real-time working condition of the pumping unit can be obtained.

4. Experiment and Analysis

In this paper, the UAV is used for video capture, and the video is processed by frame separation. The image marking tool is used to mark the pumping unit and the head working, and the training data set is produced. The Tensorflow-GPU [37] version is used as a framework for deep learning, implemented under the Linux operating system, using 1080Ti GPU for image training and target detection and tracking in the video. The detection speed and mAP value are used to analyze the advantages of the YOLOv3 algorithm as a detector in the framework proposed in this paper, to achieve a good real-time tracking effect, and make decisions for the detection of the working condition.

4.1. Description of the Training Data

In this paper, UAV aerial photography inspection data provided by China Petroleum Western Drilling Engineering Co., Ltd., were screened through screening and editing to select 5 videos for 130 min with a resolution of 640 × 480. Four of them are medium and low altitude flight (15–25 m), and one video is high altitude flight (45–55 m). After the data from three videos were processed by interval frames, the parts of the data that did not meet the training conditions were removed. The training set contained about 5400 images, and the data from the remaining two videos were processed by interval frames as the test set images, with about 2500 images. A part of the data set is shown in Figure 6.

Figure 6. Part of the image in the dataset.

After the video data was processed in a frame-by-frame process, the pumping unit and the head working were manually labeled using an image labeling tool. After each image was annotated, a class label file was generated, which stores the position of the label box and the category information, as shown in Table 1:

Table 1. The examples of annotated data.

Class ID	Normalization of the Central Point x Value	Normalization of the Central Point y Value	Normalization of w Value	Normalization of h Value
0	0.6698369565217391	0.4565916398713826	0.34148550724637683	0.9131832797427653
0	0.3675781250000003	0.6263888888888889	0.20390625	0.4666666666666667
0	0.3583333333333334	0.5857142857142857	0.46	0.5428571428571429
1	0.5083333333333334	0.4768356643356643	0.6266666666666667	0.7159090909090909
1	0.6391666666666667	0.65	0.22166666666666668	0.3666666666666667
1	0.6216666666666667	0.43214285714285716	0.41000000000000003	0.7214285714285715

It can be shown from Table 1 that the class label with ID 0 is the pumping unit, and the class label with ID 1 belongs to the head working. The center point x,y coordinate value, the width value w, and the height value h of the detection frame are all normalized according to the image size.

The purpose of frame separation processing is to improve the processing speed of the whole system without affecting the prediction ability of the Kalman filter. The direct impact of video frame separation processing is whether the target position change rate can be learned. If the interval is too long, there will be a phenomenon in which the image information is not extracted when the target trajectory changes greatly, which will lead to an unstable change of the learned position. In order to verify the effect of different interval frames on the learning ability of the frame on the target trajectory, each video in this paper trains the network at intervals of 1, 5, 10, 15, and 25 frames and tests them. Figure 7 shows the average of the loss results of the three videos of the training set after different interval frame numbers.

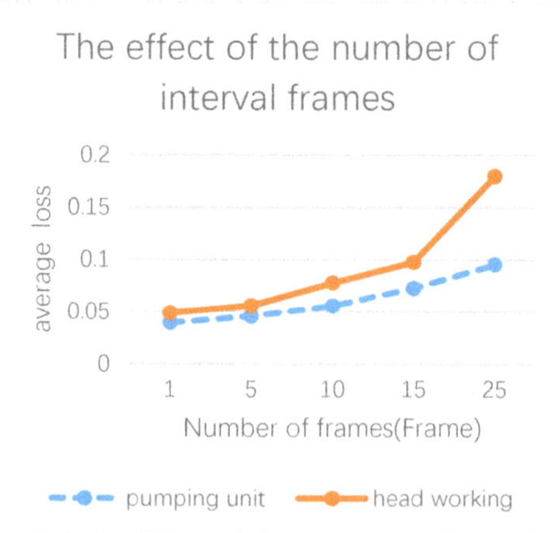

Figure 7. Comparison of average loss results at different intervals.

As can be seen from Figure 7, there is a small gap between different interval frames in the prediction of the target trajectory in the video total, especially in the case of small intervals, but when the interval frames are too large, the prediction ability will decline sharply. It can be concluded from the results that the Kalman filter can learn the target motion rule well. However, when the frame interval time is larger, the target motion regularity is weaker, and the prediction effect will be worse. Because the pumping unit is in a working state during the inspection of the drone, the head working is often obscured or incomplete. Therefore, the data of head working in the obtained data is relatively poor compared to the overall pumping unit. When the number of interval frames is large, the loss result will also be worse than that of the pumping unit.

4.2. Experimental Results and Comparison

The tracking framework used in this article makes the tracking effect dependent on the quality of the detector; therefore, different detectors are used in this paper to make comparative experiments. SSD is an algorithm similar to YOLOv3 in performance and core thinking; thus, the comparison of detectors in the following section is mainly to compare SSD with YOLOv3. Figure 8 shows the detection and tracking effect of a single target. The blue box is the detection box, and the white one is the tracking box. The purpose of the detection box is to accurately find the location and size of the target to be found in each frame and mark it out. The tracking box relies on the detection box to match the detection box before and after the frame and to predict the motion and similarity of the tracking target. For the occluded target, the detection box will not appear, because there is no target to be detected in the image. In this paper, for the occluded target in a short time, the tracking box will continue to track it according to the prediction in the previous frame.

Figure 8. Single target effects.

Whether it is SSD or YOLOv3 as a detector, the detection of a single target can get better results. Although the pumping unit has no movement change, with the movement of the UAV, the detection box and tracking box can accurately follow the target. Moreover, the box of the head working can also move as it moves up and down.

In the case of a medium or low flight height of the UAV, different algorithms are used as detectors to detect and track multiple targets, which are shown in Figures 9 and 10.

Figure 9. The tracking effect with Single Shot Multi-Box Detector (SSD) as the detector.

Figure 10. The tracking effect with YOLOv3 as the detector.

It can be shown from Figure 9 that a pumping unit in the lower left corner was not detected, which also led to the failure of tracking, while the tracking of YOLOv3 as a detector succeeded. However, neither achieved a tracking effect on targets with long-term obscuration, which is the sacrifice of the sort algorithm in this framework to achieve a faster tracking speed. However, the flexibility based on drones can make up for this shortcoming. In terms of speed, YOLOv3 as the detector is faster, which is also the advantage of the algorithm for the detecting speed. For the shadowing problem, this paper makes the following test to test the critical value of tracking failure.

As shown in Figure 11, with a test for the critical value of the tracking effect in the case of shielding, it can be seen that the four images are continuously intercepted while the head working of the left pumping unit is slowly leaving the video viewing angle. The head working in the first three pictures is still in the line of sight of the drone, but are slowly decreasing. Still, it can still be tested and tracked, and the last one shows that when the head working disappears completely in the line of sight, it immediately loses its detection and tracking effect. Moreover, there is also a pumping unit behind the pumping unit on the left side of Figures 1–3, but they are not detected and tracked because of the occlusion. This is because the sort tracking algorithm only uses the position and size of the detection frame to perform the motion estimation and data association of the target in pursuit of the tracking speed. When the target is lost, it cannot be found, and the ID can only be re-updated through detection. Therefore, the critical value in the case of occlusion is that the tracking effect is lost when the occlusion is completely occluded or the detector does not detect the target due to occlusion. However, this is when the target disappears in the entire image. In Figures 9 and 10, multiple pumping units are working side by side, which causes the pumping unit and head working to be obscured by other pumping units. When the obstacle is detected, the Kalman filter can predict the position of the object in the detection box at the next moment. However, this prediction is very rough. When the object appears again, it is tracked through matching. However, the frame proposed in this paper is exactly in line with the scene of the oil field, and the shielding time is almost zero. Moreover, the UAV is in the way of patrol inspection, which also increases the probability of avoiding shielding and reduces the shielding time.

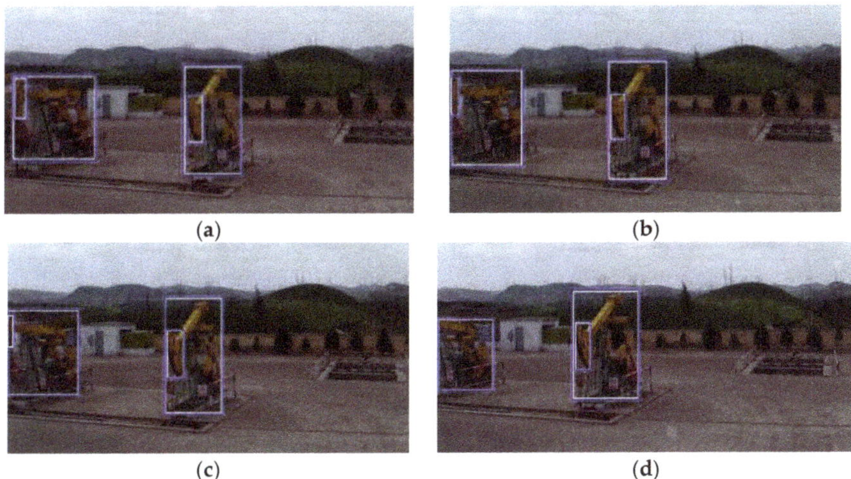

Figure 11. Masking of the critical value test. (**a**) Frame 180, (**b**) frame 185, (**c**) frame 188, (**d**) frame 196.

In the case of the UAV flying at a high altitude, different algorithms are used as detectors to detect and track multiple targets, which are shown in Figures 12 and 13.

Figure 12. The tracking effect with SSD as detector.

Figure 13. The tracking effect with YOLOv3 as detector.

It can be seen from Figures 12 and 13 that the detection and tracking effect of the pumping unit was achieved, but the tracking result with SSD as the detector did not detect and track the head working position. This is related to the performance of the detector. For YOLOv3, the defects of small targets that could not be detected in the previous series have been improved, so compared to SSD as a detector, YOLOv3 has a better effect on detecting small targets. The tracking effect in this paper also depends on the quality of the detector, so it can be seen that the tracking effect with SSD as the detector does not track small targets.

4.3. The Analysis of Experiment

When training in the detector section, the default number of iterations for YOLOv3 training is 500,200. After 500,200 iterations, the training will stop automatically. Training can also be stopped when the loss is no longer falling or the drop is very slow. The training log should be saved after the training and the following loss curve drawn using python. In order to make the contrast more vivid, the training loss curve of the SSD is drawn by taking the iteration times and the same iteration interval of YOLOv3.

As shown in Figure 14, the training stops at 16,000 iterations, and the loss value finally converges to 0.05. In this experiment, since the average loss of YOLOv3 is very slow and substantially converged after less than 0.05, the threshold for stopping the training is set to 0.05 at the time of this training. When the loss value reaches 0.05, the number of iterations is about 16,000. Therefore, the training iteration of the SSD is also taken from the log between 6000 and 16,000 to draw Figure 15. The above two loss graphs show that YOLOv3 basically converges to 0.05, and after 16,000 iterations, the SSD's loss curve still fluctuates between 0.25 and 0.1 and does not converge. It can be concluded that YOLOv3 has the advantage of training. The loss curve is not only faster than SSD convergence but also has a smaller convergence value. Therefore, YOLOv3 is more suitable as a detector in this paper than the SSD algorithm.

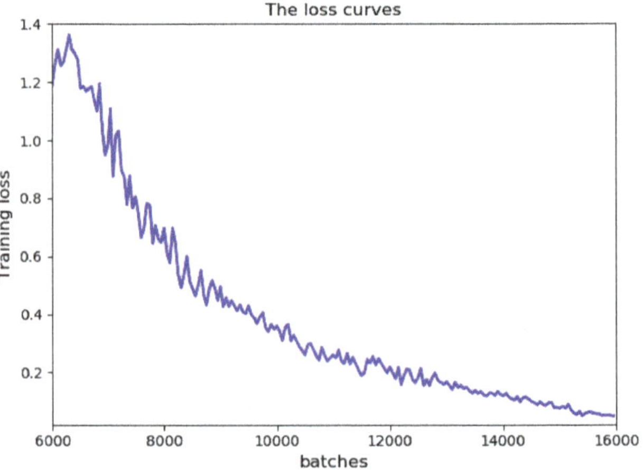

Figure 14. The loss curves of YOLOv3.

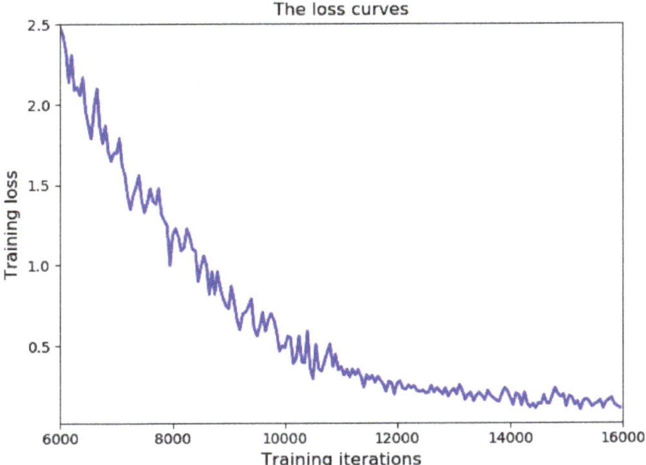

Figure 15. The loss curves of SSD.

In the experiment of this paper, the SSD algorithm with similar performance to YOLOv3 is compared with the algorithm used in this paper to compare the advantages of YOLOv3 as the detector in this paper. However, a comparison of the Faster R-CNN algorithm [38], the most advanced in object detection based on deep learning R-CNN series images, is also added. The comparison is mainly made from two aspects, the detection speed and mAP value. The former directly affects the real-time detection and tracking of the pumping unit and the head working, while the latter reflects the accuracy of detection and is the performance evaluation of the detector.

The mean average precision (mAP) is shown in Formula (7):

$$mAP = \int_0^1 P(R)d(R) \tag{7}$$

where P is the accuracy of the pumping unit and the head working, and R is their recall rate. The formulas for R and P are shown in Formulas (8) and (9), respectively:

$$P = Number\ of\ targets\ detected / The\ total\ number\ of\ detected\ detection\ frames \tag{8}$$

$$R = The\ total\ number\ of\ detected\ targets / Verify\ the\ total\ number\ of\ all\ marked \\ pumping\ units\ and\ the\ head\ working\ in\ the\ set \tag{9}$$

As shown in Table 2, mAP values and the target detection speed of the three algorithms are respectively displayed.

Table 2. Test results for the three models.

Model	mAP(%)	Time for Detection(s)
Faster R-CNN	57.6	248
SSD	64.7	39
YOLOv3	64.5	20

It can be shown from Table 2 that YOLOv3 reached 64%; although the mAP of YOLOv3 is 0.02% less than that of SSD, it is almost the same. However, in terms of time, YOLOv3 only uses 20 s, which is much shorter than the time of the above two algorithms. It fully meets the requirements of real-time performance emphasized in this paper. Therefore, it can be concluded that YOLOv3 is the most suitable detector for this experiment in terms of both accuracy and speed.

Finally, we compare the advantages and disadvantages of the proposed framework with other multi-target tracking algorithms, as this paper focuses on industrial applications, especially in this paper, for the tracking of oil field pumping units and head working. Therefore, the first two methods with the fastest processing speed of MOT Challenge2016 are selected for comparison. According to the size of the MOTA scores, the comparison results are shown in Table 3.

Table 3. The quality of evaluation of different methods.

Tracker	MOTA	MOPI	FP	FN	ID SW	HZ
YLTS	57.6	79.6	8698	63,245	1423	60.1
SMMUML	43.3	74.8	8463	93,892	985	187.2
LP2D	35.7	75.5	5084	111,163	1264	49.3

As shown in Table 3, the two algorithms with the fastest processing speed are compared with the framework proposed in this paper. The fastest algorithm is 182.7 HZ, which is far higher than all other algorithms. The processing speed of the framework proposed in this paper ranks second, which is more suitable for industrial applications, thanks to the processing speed of YOLOv3 and the sort algorithm. However, some other factors are sacrificed. IDSW is relatively high, which is also used to improve the speed and lead to more ID changes. Generally speaking, this framework achieves the second level in terms of processing speed on the premise of maintaining a high MOTA level. In combination with speed and accuracy, it can be seen that the proposed multi-target tracking framework has achieved good results.

5. Conclusions and Future Works

The Faster R-CNN, SSD, and YOLOv3 algorithms used in the experiments in this paper were used as detectors in the tracking framework proposed in this paper. The framework uses sort tracking to meet the real-time nature of the oilfield well conditions, which also puts the focus of this framework on the detector. The quality of the tracking depends entirely on the quality of the detector. Experiments have shown that YOLOv3 is the most suitable detector for this article, both in terms of accuracy and speed. However, the framework also has shortcomings. The detector and tracker used in this paper are designed to meet real-time performance, so it is faster in speed, but it also sacrifices tracking in special cases. For example, in the case of a long-term occlusion, the target being tracked will be lost, and the target ID will be frequently switched, which reduces the tracking effect. However, based on the background of the drone's refined inspection, this situation has also been reduced. Therefore, the final result can be used to track the pumping unit and the key components, such as head working, to obtain the position and motion information of the target, and to provide a basis for further semantic layer analysis (motion state recognition, scene recognition, etc.). In this way, the working conditions are checked in real-time.

According to the current research results, this paper believes that although the tracking target does not appear to be occluded for too long in the scene of drone inspection, it cannot ignore the existence of this situation. Considering the problem of target occlusion in the tracker is a concern for future research. This also reduces the dependence on the detector, reduces the number of ID switching during the tracking target, and improves the overall tracking performance. After obtaining the information of the tracking target, further motion analysis of the target working state to obtain clearer working conditions is also a concern for future research.

Author Contributions: Conceptualization, Y.Z., C.W., Q.W. and N.X.; data curation, Y.Z., Q.W. and Z.M.E.; formal analysis, Y.Z., C.W., N.X., Z.M.E. and S.Z.; funding acquisition, C.W.; investigation, Y.Z., Q.W. and S.Z.; methodology, Y.Z.; project administration, C.W., N.X. and S.Z.; resources, C.W., Q.W. and Z.M.E.; software, Y.Z.

and S.Z.; supervision, C.W., N.X. and S.Z.; validation, Y.Z.; visualization, Y.Z., Q.W. and Z.M.E.; writing—original draft, Y.Z.; writing—review & editing, Y.Z., C.W. and N.X.

Funding: This research was supported by Technology Innovation Action Plan Project (19511105103, 17511107203) and the National Key Research and Development Program of China (2018YFC0810204, 2018YFB17026) and National Natural Science Foundation of China (61872242), Shanghai Science and Shanghai key lab of modern optical system.

Acknowledgments: The authors would like to appreciate all anonymous reviewers for their insightful comments and constructive suggestions to polish this paper in high quality. Thanks to the data provided by China Petroleum West Drilling Engineering Co., Ltd. to support this research.

Conflicts of Interest: The authors declare no conflict of interest.

References

1. Chen, H.; Wu, C.; Huang, W.; Wu, Y.; Xiong, N. Design and Application of System with Dual-Control of Water and Electricity Based on Wireless Sensor Network and Video Recognition Technology. *Int. J. Distrib. Sens. Netw.* **2018**, *14*, 1550147718795951. [CrossRef]
2. Li, X.; Zhou, C.; Tian, Y.-C.; Xiong, N.; Qin, Y. Asset-Based Dynamic Impact Assessment of Cyberattacks for Risk Analysis in Industrial Control Systems. *IEEE Trans. Ind. Inform.* **2018**, *14*, 608–618. [CrossRef]
3. Ju, C.; Son, H. Multiple Uav Systems for Agricultural Applications: Control, Implementation, and Evaluation. *Electronics* **2018**, *7*, 162. [CrossRef]
4. Hu, G.; Yang, Z.; Han, J.; Huang, L.; Gong, J.; Xiong, N. Aircraft Detection in Remote Sensing Images Based on Saliency and Convolution Neural Network. *EURASIP J. Wirel. Comm. Netw.* **2018**, *2018*, 26. [CrossRef]
5. Hua, X.; Wang, X.; Rui, T.; Wang, D.; Shao, F. Real-Time Object Detection in Remote Sensing Images Based on Visual Perception and Memory Reasoning. *Electronics* **2019**, *8*, 1151. [CrossRef]
6. Aksu, D.; Aydin, M.A. Detecting Port Scan Attempts with Comparative Analysis of Deep Learning and Support Vector Machine Algorithms. In Proceedings of the 2018 International Congress on Big Data, Deep Learning and Fighting Cyber Terrorism (IBIGDELFT), Ankara, Turkey, 3–4 December 2018. [CrossRef]
7. Wu, C.; Luo, C.; Xiong, N.; Zhang, W.; Kim, T. A Greedy Deep Learning Method for Medical Disease Analysis. *IEEE Access* **2018**, *6*, 20021–20030. [CrossRef]
8. Huang, H.; Xu, Y.; Huang, Y.; Yang, Q.; Zhou, Z. Pedestrian Tracking by Learning Deep Features. *J. Vis. Commun. Image Represent.* **2018**, *57*, 172–175. [CrossRef]
9. Wang, S.; Wu, C.; Gao, L.; Yao, Y. Research on Consistency Maintenance of the Real-Time Image Editing System Based on Bitmap. In Proceedings of the 2014 IEEE 18th International Conference on Computer Supported Cooperative Work in Design (CSCWD), Hsinchu, Taiwan, 21–23 May 2014. [CrossRef]
10. Zhao, J.; Xu, H.; Liu, H.; Wu, J.; Zheng, Y.; Wu, D. Detection and Tracking of Pedestrians and Vehicles Using Roadside Lidar Sensors. *Transp. Res. Part C Emerg. Technol.* **2019**, *100*, 68–87. [CrossRef]
11. Jiang, X.; Fang, Z.; Xiong, N.N.; Gao, Y.; Huang, B.; Zhang, J.; Yu, L.; Harrington, P. Data Fusion-Based Multi-Object Tracking for Unconstrained Visual Sensor Networks. *IEEE Access* **2018**, *6*, 13716–13728. [CrossRef]
12. Liu, C.; Zhou, A.; Wu, C.; Zhang, G. Image Segmentation Framework Based on Multiple Feature Spaces. *IET Image Process.* **2015**, *9*, 271–279. [CrossRef]
13. Yang, J.C.; Jiao, Y.; Xiong, N.; Park, D.S. Fast Face Gender Recognition by Using Local Ternary Pattern and Extreme Learning Machine. *TIIS* **2013**, *7*, 1705–1720.
14. Xue, W.; Wenxia, X.; Guodong, L. Image Edge Detection Algorithm Research Based on the Cnns Neighborhood Radius Equals 2. In Proceedings of the 2016 International Conference on Smart Grid and Electrical Automation (ICSGEA), Zhangjiajie, China, 11–12 August 2016. [CrossRef]
15. Lecun, Y.; Bottou, L.; Bengio, Y.; Haffner, P. Gradient-Based Learning Applied to Document Recognition. *Proc. IEEE* **1998**, *86*, 2278–2324. [CrossRef]
16. Wan, L.D.; Zeiler, M.; Zhang, S.; Lecun, Y.; Fergus, R. Regularization of Neural Networks Using Dropconnect. *Int. Conf. Mach. Learn.* **2013**, *28*, 1058–1066.
17. Wang, Z.; Lu, W.; He, Y.; Xiong, N.; Wei, J. *Re-Cnn: A Robust Convolutional Neural Networks for Image Recognition*; Springer: Berlin, Germany, 2018.

18. Napiorkowska, M.; Petit, D.; Marti, P. Three Applications of Deep Learning Algorithms for Object Detection in Satellite Imagery. In Proceedings of the IGARSS 2018-2018 IEEE International Geoscience and Remote Sensing Symposium, Valencia, Spain, 22–27 July 2018. [CrossRef]

19. Ren, S.; He, K.; Girshick, R.; Sun, J. Faster R-Cnn: Towards Real-Time Object Detection with Region Proposal Networks. *IEEE Trans. Pattern Anal. Mach. Intell.* **2017**, *39*, 1137–1149. [CrossRef]

20. Lan, W.; Dang, J.; Wang, Y.; Wang, S. Pedestrian Detection Based on Yolo Network Model. In Proceedings of the 2018 IEEE International Conference on Mechatronics and Automation (ICMA), Changchun, China, 5–8 August 2018. [CrossRef]

21. Wang, X.; Hua, X.; Xiao, F.; Li, Y.; Hu, X.; Sun, P. Multi-Object Detection in Traffic Scenes Based on Improved Ssd. *Electronics* **2018**, *7*, 302. [CrossRef]

22. Biswas, D.; Su, H.; Wang, C.; Stevanovic, A.; Wang, W. An Automatic Traffic Density Estimation Using Single Shot Detection (Ssd) and Mobilenet-Ssd. *Phys. Chem. Earth Parts A/B/C* **2018**, *110*, 176–184. [CrossRef]

23. Kitayama, T.; Lu, H.; Li, Y.; Kim, H. Detection of Grasping Position from Video Images Based on Ssd. In Proceedings of the 2018 18th International Conference on Control, Automation and Systems (ICCAS), Daegwallyeong, Korea, 17–20 October 2018.

24. Zhang, H.; Gao, L.; Xu, M.; Wang, Y. An Improved Probability Hypothesis Density Filter for Multi-Target Tracking. *Optik* **2019**, *182*, 23–31. [CrossRef]

25. Yang, T.; Cappelle, C.; Ruichek, Y.; El Bagdouri, M. Online Multi-Object Tracking Combining Optical Flow and Compressive Tracking in Markov Decision Process. *J. Vis. Commun. Image Represent.* **2019**, *58*, 178–186. [CrossRef]

26. Shinde, S.; Kothari, A.; Gupta, V. Yolo Based Human Action Recognition and Localization. *Proced. Comput. Sci.* **2018**, *133*, 831–838. [CrossRef]

27. Krawczyk, Z.; Starzyński, J. Bones Detection in the Pelvic Area on the Basis of Yolo Neural Network. In Proceedings of the 19th International Conference Computational Problems of Electrical Engineering, Banska Stiavnica, Slovakia, 9–12 September 2018. [CrossRef]

28. Liu, X.; Yang, T.; Li, J. Real-Time Ground Vehicle Detection in Aerial Infrared Imagery Based on Convolutional Neural Network. *Electronics* **2018**, *7*, 78. [CrossRef]

29. Tian, Y.; Yang, G.; Wang, Z.; Wang, H.; Li, E.; Liang, Z. Apple Detection During Different Growth Stages in Orchards Using the Improved Yolo-V3 Model. *Comput. Electr. Agric.* **2019**, *157*, 417–426. [CrossRef]

30. Tumas, P.; Serackis, A. Automated Image Annotation Based on Yolov3. In Proceedings of the 2018 IEEE 6th Workshop on Advances in Information, Electronic and Electrical Engineering (AIEEE), Vilnius, Lithuania, 8–10 November 2018. [CrossRef]

31. Huang, R.; Gu, J.; Sun, X.; Hou, Y.; Uddin, S. A Rapid Recognition Method for Electronic Components Based on the Improved Yolo-V3 Network. *Electronics* **2019**, *8*, 825. [CrossRef]

32. Shi, Z.; Xu, X. Near and Supersonic Target Tracking Algorithm Based on Adaptive Kalman Filter. In Proceedings of the 2016 8th International Conference on Intelligent Human-Machine Systems and Cybernetics (IHMSC), Hangzhou, China, 27–28 August 2016. [CrossRef]

33. Reza, Z.; Buehrer, R.M. *An Introduction to Kalman Filtering Implementation for Localization and Tracking Applications*; The Institute of Electrical and Electronics Engineers, Inc.: New York, NY, USA, 2018.

34. Liu, Y.; Wang, P.; Wang, H. Target Tracking Algorithm Based on Deep Learning and Multi-Video Monitoring. In Proceedings of the 2018 5th International Conference on Systems and Informatics (ICSAI), Nanjing, China, 10–12 November 2018. [CrossRef]

35. Li, H.; Qin, J.; Xiang, X.; Pan, L.; Ma, W.; Xiong, N.N. An Efficient Image Matching Algorithm Based on Adaptive Threshold and Ransac. *IEEE Access* **2018**, *6*, 66963–66971. [CrossRef]

36. Tounsi, K.; Abdelkader, D.; Iqbal, A.; Sanjeevikumar, P.; Barkat, S. Extended Kalman Filter Based Sliding Mode Control of Parallel-Connected Two Five-Phase Pmsm Drive System. *Electronics* **2018**, *7*, 14.

37. Demirović, D.; Skejić, E.; Šerifović–Trbalić, A. Performance of Some Image Processing Algorithms in Tensorflow. In Proceedings of the 2018 25th International Conference on Systems, Signals and Image Processing (IWSSIP), Maribor, Slovenia, 20–22 June 2018. [CrossRef]
38. Beibei, Z.; Xiaoyu, W.; Lei, Y.; Yinghua, S.; Linglin, W. Automatic Detection of Books Based on Faster R-Cnn. In Proceedings of the 2016 Third International Conference on Digital Information Processing, Data Mining, and Wireless Communications (DIPDMWC), Moscow, Russia, 6–8 July 2016. [CrossRef]

 *electronics* 

Article

Accurate Landing of Unmanned Aerial Vehicles Using Ground Pattern Recognition

Jamie Wubben [1], Francisco Fabra [2], Carlos T. Calafate [2,*], Tomasz Krzeszowski [3], Johann M. Marquez-Barja [1,4], Juan-Carlos Cano [2] and Pietro Manzoni [2]

[1]   Faculty of Applied Engineering, Electronics-ICT, IDLab University of Antwerp, 2000 Antwerp, Belgium; jamie.wubben@student.uantwerpen.be (J.W.); Johann.Marquez-Barja@uantwerpen.be (J.M.M.-B.)

[2]   Departament of Computer Engineering (DISCA), Universitat Politècnica de València, 46022 Valencia, Spain; frafabco@cam.upv.es (F.F.); jucano@disca.upv.es (J.-C.C.); pmanzoni@disca.upv.es (P.M.)

[3]   Faculty of Electrical and Computer Engineering, Rzeszow University of Technology, 35-959 Rzeszow, Poland; tkrzeszo@prz.edu.pl

[4]   imec Antwerp, 2000 Antwerpen, Belgium

*   Correspondence: calafate@disca.upv.es

Received: 6 November 2019; Accepted: 3 December 2019; Published: 12 December 2019

Abstract: Over the last few years, several researchers have been developing protocols and applications in order to autonomously land unmanned aerial vehicles (UAVs). However, most of the proposed protocols rely on expensive equipment or do not satisfy the high precision needs of some UAV applications such as package retrieval and delivery or the compact landing of UAV swarms. Therefore, in this work, a solution for high precision landing based on the use of ArUco markers is presented. In the proposed solution, a UAV equipped with a low-cost camera is able to detect ArUco markers sized 56×56 cm from an altitude of up to 30 m. Once the marker is detected, the UAV changes its flight behavior in order to land on the exact position where the marker is located. The proposal was evaluated and validated using both the ArduSim simulation platform and real UAV flights. The results show an average offset of only 11 cm from the target position, which vastly improves the landing accuracy compared to the traditional GPS-based landing, which typically deviates from the intended target by 1 to 3 m.

Keywords: UAV; autonomous landing; vision-based; ArduSim; ArUco marker

---

1. Introduction

Recently there has been a growing interest in unmanned aerial vehicles (UAVs). Their applications are diverse, ranging from surveillance, inspection and monitoring, to precision agriculture and package retrieval/delivery. Landing a UAV is acknowledged as the last and most critical stage of navigation [1]. According to statistics, the number of accidents associated to the UAV landing process represent 80% of the hazard cases [2]. Therefore, improved landing techniques are being intensely explored. Furthermore, some of the applications mentioned above, such as package retrieval, require a high level of accuracy so as to make sure the UAV lands exactly on the desired target.

Previous proposals heavily rely on the Global Positioning System (GPS) and inertial navigation sensors (INS) as the main positioning approaches [3]. However, altitude data provided by the GPS is typically inaccurate and needs to be compensated with a close-range sensor, such as a barometric pressure sensor or a radar altimeter [4]. Despite such compensation, these methods still remain inaccurate, especially in the horizontal plane, resulting in a landing position that typically deviates from the intended one by 1 to 3 m. Furthermore, the GPS cannot be used indoors. For these reasons, GPS and INS systems are mostly used for long range, outdoor flights having low accuracy requirements [3].

Taking the aforementioned issues into consideration, the aim of this work is to develop a novel vision-based landing system that is able to make a UAV land in a very specific place with high precision. This challenge is addressed by developing a solution that combines the use of a camera and ArUco markers [5,6]. This way, the relative offset of the UAV towards the target landing position is calculated using the ArUco library [7] (based on OpenCV). After computing its relative offset, the UAV adjusts its position so as to move towards the center of the marker and start descending, performing additional adjustments dynamically.

The rest of this work is structured as follows: in the next section some related works on UAV landing strategies are presented. In Section 3, our proposed solution is introduced, detailing the methodology followed in order to track the target for landing and how to perform the necessary calculations to adjust the UAV position. Then, Section 4 provides some technical details about the UAV used to deploy our proposal, highlighting the main issues and restrictions to be taken into consideration in the design of our novel landing algorithm. Section 5 describes how the different experiments were made. The main results are then presented in Section 6, where our solution is compared to a GPS-based approach, with an appropriate discussion. Finally, Section 7 concludes this work and refers to future works.

2. Related Work

Recently, different UAV landing approaches have been studied. These studies can be categorized based on the different types of landing platforms adopted. According to Reference [8], the vast amount of landing platforms belong to one of these three categories:

1. *Category I–fixed platforms:*
    Fixed platforms include all platforms that are stationary. This type of platforms are the easiest to land on, since the target remains stationary.
2. *Category II–moving platforms:*
    Moving platforms are defined by the ability to move with two degrees (surge, sway) of freedom. In this case, the UAV needs to track the platform first and then land on it.
3. *Category III–Landing on a ship:*
    The ultimate landing platform is on a moving ship. In this case the ship has six degrees of freedom (heave, sway, surge, yaw, roll, pitch) and, therefore, developing a safe landing approach becomes a troublesome task.

It is also possible to categorize the different landing approaches based on the sensor(s) used in the process. Many different sensors can be used, such as sonar, infrared, LIDAR, cameras or a combination of these. Below, some approaches that rely on computer vision will be discussed.

Chen et al. [9] succeeded in landing a real UAV on an object moving with a speed of 1 m/s (category II). A camera was used to track the position of the landing platform (xy-coordinates) and a LIDAR sensor provided detailed information about the altitude. This research work introduced a robust method to track and land on a moving object. However, the use of a LIDAR sensor discourages the solution, as it tends to be too expensive when scaling to a high number of UAVs.

Nowak et al. [10] proposed a system in which a UAV could land both at night, as well as during the day. The idea is simple yet robust and elegant: a beacon is placed on the ground. The light emitted from the beacon is then captured by a camera (without a infrared filter) and the drone moves (in the xy-plane) such that the beacon is in the center of the picture. Once centerd, the height is estimated based on the image area occupied by the beacon and the drone's altitude is decreased in order for it to land safely.

Shaker et al. [11] suggested another approach: reinforcement learning. In this approach, the UAV agent learns and adapts its behaviour when required. Usually, reinforcement learning takes a lot of time. To accelerate this process, a technique called Least-Squares Policy Iteration (LSPI) is used. With this method, a simulated UAV (AR100) was able to achieve a smooth landing trajectory swiftly.

In the work of Lange et al. [12] an approach for landing and position control, similar to our work, was developed. Their approach was also based on OpenCV and on recognizing a landing pattern. However, their landing pattern was not built with the use of ArUco markers. In fact, the landing pattern used, with a diameter of 45 cm, was only detected from a distance of 70 cm. Therefore, this strategy cannot be used in an outdoor environment where the flight altitude is typically much higher. However, in this approach, the UAV does not need to see the entire marker, which is an advantage of this scheme.

A system that can land on and track a slow moving vehicle (180 cm/s) was developed by Araar et al. [13]. Indoor experiments show that the UAV used was able to successfully land on the target landing platform (which also consists of multiple ArUco markers) from a height of approximately 80 cm.

More recently, Patruno et al. [14] presented a solution for the landing of UAVs on a human-made landing target. Their target was similar to traditional heliplatforms but with specific aspect ratios, so that it can be detected from long distances. The geometric properties of the H-shaped marks adopted are used to estimate the pose with high accuracy, achieving an average RMSE value of only 0.0137 m in pose and $1.04°$ in orientation.

Baca et al. [15] were able to detect a moving car at 15 km/h, predict its future movement and attach to it. To achieve it they equipped the UAV with onboard sensors and a computer, which detects the car using a monocular camera and predicts the car future movement using a nonlinear motion model. While following the car, the UAV lands on its roof and it attaches itself using magnetic legs.

De Souza et al. [16] developed a autonomous landing system based on Artificial Neural Network (ANN) supervised by Fuzzy Mamdani Logic. Their method introduced low computational complexity while maintaining the characteristics and intelligence of the fuzzy logic controller. They validated their solution using both simulation and real tests for static and dynamic landing spots.

Fraczek et al. [17] presented an embedded video system that allows the UAV to automatically detect safe landing sites. Their solution was implemented on a heterogeneous Zynq SoC device from Xilinx and a Jetson GPU. Differently from the previous works, this work does not rely on a human made marker. Through the use of machine learning and computer vision techniques, the UAV classifies the terrain into three classes. The proposed solution was tested on 100 test images and classified the different terrains correctly in 80% of the cases. Furthermore, in these tests the performance between the Zynq SoC device and the Jetson GPU was compared.

Our work differs from the former ones as we want the UAV to detect the landing area when it is high above the ground (height > 20 m) to compensate for possibly high GPS error values, while using cheap sensors (only a Raspberry Pi camera is needed) and yet still achieving very low errors in terms of landing accuracy (<20 cm). To that extent, only one other source (the Ardupilot community [18]) was found that is attempting to achieve results similar to ours. Their method follows the same strategy as the work of Nowak et al. [10], as they also make use of an IR-beacon. According to the ArduPilot authors, their proposed method is able to land a UAV from an altitude of 15 m, reliably under all lighting conditions and with an maximum offset of only 30 cm. While the results of this approach are impressive, ours still outperforms it in terms of both accuracy, altitude and price.

3. Proposed Solution

The aim of this work is to make a UAV land on a specific location. First the UAV has to make a coarse approach to the landing zone. As stated before, the UAV will typically fail to hover above its exact target location, being usually within 1 to 3 m away from the intended landing position. Once the UAV is close to the target location our protocol is activated. The first step deals with finding the marker. The ArUco marker library [5,6] (based on OpenCV) provides a function which takes the camera feed and returns information about the marker(s). ArUco markers resemble the well-known QR-codes. They carry less information than the latter ones (only an id), which makes them easier to detect. A typical ArUco marker consists of a black border and a 6x6 square of black and white smaller squares. There are different types of configurations (e.g., $3 \times 3, 4 \times 4, 7 \times 7$) known as dictionaries. A marker from a dictionary with less squares is of course easier to detect but only a small number of ids can be provided. In this work, dictionary "DICT_6X6_250" is used. As the name suggests, it provides 250 different ids, which is more than enough for our purposes. The second part consists of descending the UAV while trying to keep it centered over the marker.

In order to detect a marker two conditions must be met: (i) the marker must be fully inside the picture and (ii) each square must be uniquely identified (black or white). In this application, it is possible that the two conditions are not simultaneously met in some cases. For instance, when the drone is at a low altitude (i.e., 0.5 m) the marker is too big to fit inside the field of view of the camera; in addition, the shadow of the drone may "corrupt" the image. On the other hand, when the drone is flying at a higher altitude (e.g., 12 m) the image may be too small to be detected. Therefore, a strategy was developed that combines markers of different sizes (see Figure 1), so that the drone can find its target from a higher altitude. If the UAV is able to detect a smaller marker, it will switch to it, adjusting its course accordingly. Figure 2 shows a real scenario where the UAV is able to see two markers but chooses to move towards the smaller marker. The center of this marker is indicated by the red spot.

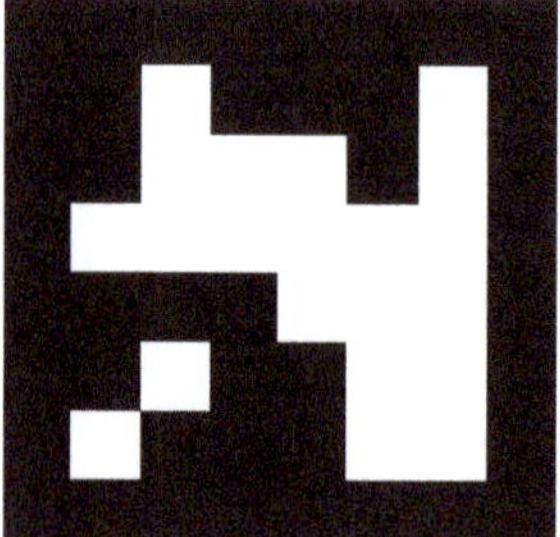
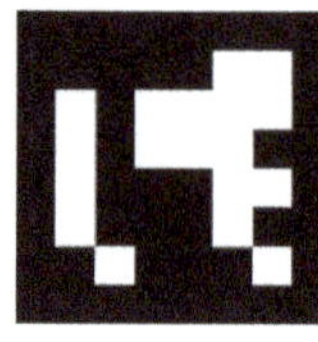

Figure 1. Two examples of ArUco markers of different sizes.

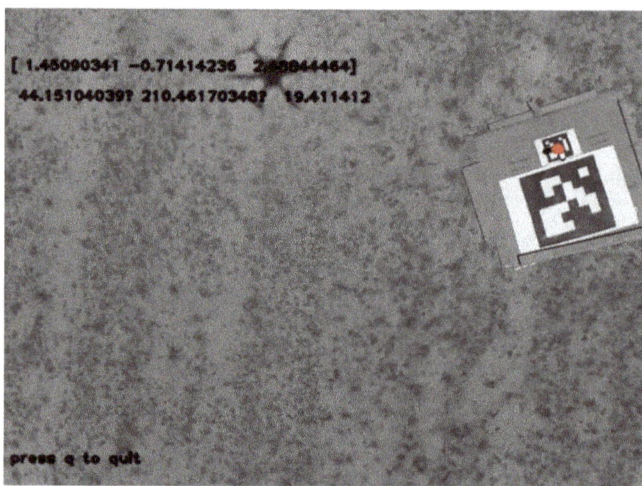

Figure 2. Image retrieved by the Unmanned Aerial Vehicle (UAV) camera after processing using OpenCV/ArUco libraries.

Once the marker is detected, the drone has to move towards the center of the marker and descend from there. Due to the effect of wind and to the inherent instability of the UAV itself, the drone will also move in the horizontal plane while descending. This unwanted movement should be compensated in order to land the drone more precisely. To achieve this behavior, the strategy described in Algorithm 1 is proposed, which works as follows: In line 3, the UAV searches for an ArUco marker. If no marker is detected, the flight mode of the UAV is changed to loiter. If this is the case for 30 consecutive seconds, the mission is aborted and the UAV will land using GPS only. Otherwise, from the potential list of detected markers, the marker with the highest ID (i.e., the smallest marker) is selected (line 11). With the use of the ArUco library, the location of the marker with respect to the drone is estimated. If the altitude of the UAV is greater than z_2 (see empirical values in Table 1), α is set to 20 degrees; otherwise, it is set to 10 degrees. These values are based on: the detection distance of the markers, size of the UAV, size of the markers and a margin which is optimized empirically. In line 20 it is checked if the marker is within the virtual border (explained later). If so, the UAV descends; otherwise, it moves horizontally towards the target position. This algorithm will be executed continuously as long as the altitude of the UAV is greater than z_1. From the moment the UAV's altitude drops below z_1 (very near to ground), the control will be handed over to the flight controller, which will land the UAV in a safe manner and disarm the engines.

Algorithm 1 Static vision-based landing strategy.

1: Start timer 30 s
2: **while** altitude $> z_1$ **do**
3:     IDs, detected $\leftarrow$ SearchMarker()
4:     **if** $\neg$ detected **then**
5:         Loiter()
6:         **if** timer exceeded **then**
7:             AbortLanding()
8:         **end if**
9:     **else**
10:         reset timer()
11:         ID $\leftarrow$ highest detected ID
12:         Get $P(x, y, z)_{id}$
13:         **if** $z > z_2$ **then**
14:             $\alpha = 20°$
15:         **else**
16:             $\alpha = 10°$
17:         **end if**
18:         $\beta_x = |\arctan(x/z)|$
19:         $\beta_y = |\arctan(y/z)|$
20:         **if** $\beta_x > \alpha$ **or** $\beta_y > \alpha$ **then**
21:             Move(x,y)
22:         **else**
23:             Descend()
24:         **end if**
25:     **end if**
26: **end while**
27: Descend and disarm UAV

Table 1. Parameter values adopted regarding Algorithms 1 and 2.

Altitude threshold z_1	0.30 m
Altitude threshold z_2	13 m
Virtual border angle α	$\{10°, 20°\}$

In the description above, a virtual border is mentioned. This border defines an area which should enclose the marker (illustrated in Figure 3). This virtual border is created in order to distinguish the two cases of descending or moving horizontally. The size a of this square is defined as:

$$a = 2 \times \tan(\alpha) \times h$$

where h refers to the relative altitude of the UAV.

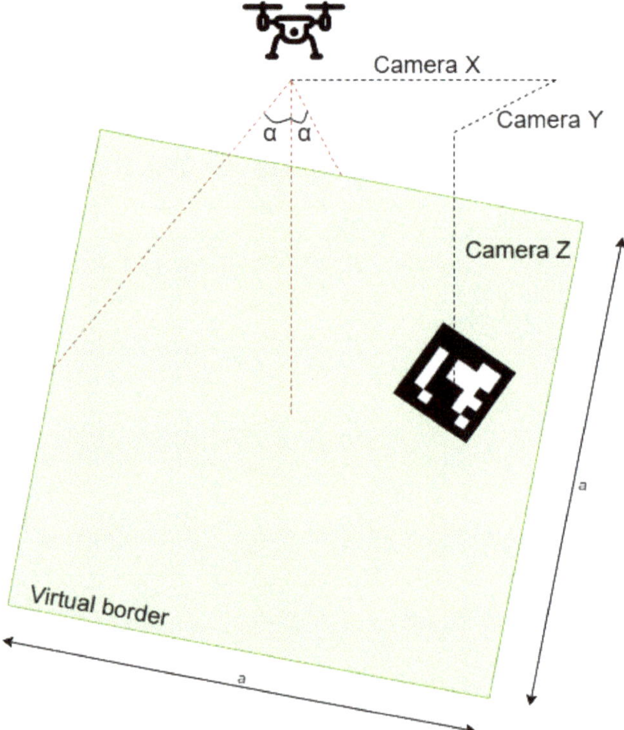

Figure 3. Visual representation of the virtual border.

The main advantage of defining the area in this way is that it will decrease as the drone lowers its altitude. Therefore, the drone will be more centerd above its target position when it flies at low altitude. However, when flying at higher altitude, the drone should descend whenever possible to avoid excessive landing times. For this reason, α is increased to $20°$ if the UAV is flying above 13 m (z_2).

There are multiple ways to move the UAV towards the target point. In ArduSim, a UAV can be moved for example, by overriding the remote control with function *"channelOverride"*. With the use of this function the UAV can be moved at a constant speed, first along the roll axis (left/right) and afterwards along the pitch (forward/backward) axis, until the target point is met.

This algorithm experiences difficulties tracking a marker, because it will quickly change between marker IDs whenever one of the markers is not visible for a short amount of time. At first, a small timer was used to eliminate this problem. However, this approach was not satisfactory and so an extension of this algorithm is proposed as Algorithm 2. This second version uses the same general ideas but introduces improvements such as hysteresis that make adjustments more dynamic, achieving a higher effectiveness than the former one. In particular, for this second version, the altitude (*activationLevel[i]*) is saved whenever a new marker is detected. Contrary to the former version, the algorithm will only switch between markers when its current altitude is lower than half of the *activationLevel[i]* value. With this modification, the typical glitching behaviour (of detecting and not detecting a marker) is bypassed. Furthermore, if the UAV is not able to detect a marker, it will switch to a recovery mode. This means that it will increase its altitude by one meter, thereby increasing the chances of finding the marker again. Finally, the horizontal speed of the UAV depends on the horizontal distance to the marker. The pitch and roll values are set to v_1 (see empirical values in Table 2) if the distance between marker and UAV is greater then 1 m (see Table 1); otherwise, it is set to v_2.

Algorithm 2 Adaptive vision-based landing strategy.

1: Start timer 30 s
2: **while** altitude $> z_1$ **do**
3:　　IDs $\leftarrow$ Search
4:　　**if** $\neg$ detected **then**
5:　　　　Recover
6:　　　　**if** timer exceeded **then**
7:　　　　　　Abort
8:　　　　**end if**
9:　　**else**
10:　　　　reset timer
11:　　　　**for all** IDs **do**
12:　　　　　　**if** First time detected **then**
13:　　　　　　　　activationLevel[i] = altitude
14:　　　　　　**end if**
15:　　　　　　**if** altitude $<$ activationLevel[i]/2 **then**
16:　　　　　　　　id $\leftarrow$ i
17:　　　　　　**end if**
18:　　　　**end for**
19:　　　　Get $P(x, y, z)_{id}$
20:　　　　**if** $z > z_2$ **then**
21:　　　　　　$\alpha = 20°$
22:　　　　**else**
23:　　　　　　$\alpha = 10°$
24:　　　　**end if**
25:　　　　$\beta_x = |\arctan(x/z)|$
26:　　　　$\beta_y = |\arctan(y/z)|$
27:　　　　**if** $\beta_x > \alpha$ **or** $\beta_y > \alpha$ **then**
28:　　　　　　Move(x,y)
29:　　　　**else**
30:　　　　　　Descend(speed)
31:　　　　**end if**
32:　　**end if**
33: **end while**
34: Descend and disarm UAV

Table 2. Speed values adopted regarding Algorithm 2.

speed v_1	15%
speed v_2	5%
speed v_3	10%

Since the area captured by the camera is large when flying at high altitudes, there is less risk of missing the marker. Therefore, the UAV's descending speed can be varied with respect to its altitude in the following Equation (1):

$$descending\ speed = \begin{cases} min\ \{60\%, altitude \times 2\%\}\ if\ altitude > 6\ meters \\ v_3\ otherwise. \end{cases} \tag{1}$$

4. UAV Specification

Due to the vast amount and diversity of UAV models available, it is worth mentioning the actual characteristics of the UAV used for our experiments. The UAV adopted belongs to the Vertical Take-Off and Landing (VTOL) category, more commonly known as a multirotor UAV. In the experiments described in this work, a hexacopter model is used (see Figure 4), being equipped with a remote control operating in the 2.4 GHz band, a telemetry channel in the 433 MHz band, a GPS receiver, a Pixhawk flight controller and a Raspberry Pi with external camera (see Figure 5). The Raspberry Pi creates an ad-hoc WiFi connection in the 5 GHz band, which is used to communicate with a ground station or with other UAVs. Below we detail the purpose and the connections (as shown in Figure 6) between these different devices.

4.1. 2.4 GHz Remote Control

The FrSky X8r receiver provides communication between the UAV and the remote control. This device makes it possible to fly the UAV manually. It receives signals from the remote control and passes them to the flight controller. Furthermore, it sends basic information about the UAV to the remote control for example, flight mode, so that the pilot is informed about the UAV state. It accomplishes these tasks by using the entire 2.4 GHz ISM band. Therefore, it becomes nearly impossible to receive/send any WiFi signal in this band (2.4 GHz), as shown in Reference [19]. Hence, our ad-hoc WiFi connections rely on the 5 GHz band instead. This 5 GHz WiFi connection is used to connect an UAV or multiple UAVs to the ArduSim ground station.

Figure 4. Hexacopter used in our experiments.

Figure 5. Raspberry Pi camera attached to the UAV.

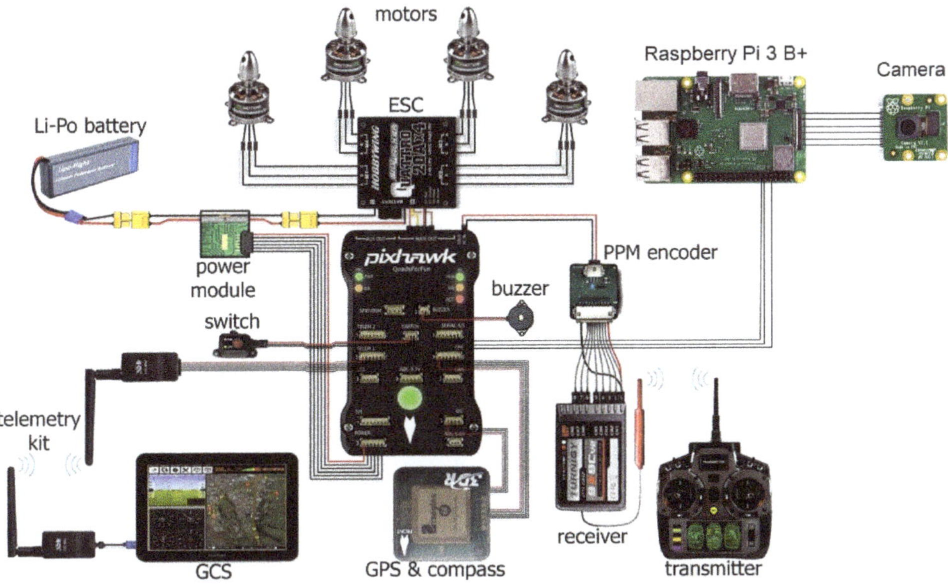

Figure 6. Wiring scheme of our UAV.

4.2. 433 MHz Telemetry

The telemetry channel operates at a lower frequency (433 MHz). Its purpose is sending UAV information from the flight controller to a ground station (typically a smartphone), including data such as heading, tilt, speed, battery lifetime, flight mode, altitude, and so forth. It is also able to receive instructions to follow a specific mission, that is, return to home or to perform an emergency landing. However, these set of instructions are very limited since the 433 MHz telemetry is downlink focused. For this reason we have chosen to use the above mentioned 5 GHz WiFi connection to control the UAV from ArduSim. The 433 MHz telemetry will only be used in case of emergency.

4.3. Flight Controller

The flight controller, in this case the Pixhawk 4, is a device that receives information from sensors and that processes this information in order to control the UAV at a low-level, while a Raspberry Pi is in charge of application level control. In particular, information from different sensors such as GPS, barometer, magnetometer, accelerometer and gyroscope are combined in order to provide an accurate representation of the UAV state. This information is then used in order to stabilize the UAV and make it controllable. Some of this information can also be sent via a serial link towards the Raspberry Pi. We opted to use the Pixhawk flight controller (containing an implementation of the Ardupilot firmware [18]) because it is an open source solution and thus new protocols are easily implemented and deployed on real UAVs. For the communication between the Raspberry Pi and the Pixhawk, the open source MAVLink protocol [20] is used. It is a lightweight messaging protocol for communicating with most open-source flight controllers, as is the case of the Pixhawk.

4.4. Raspberry Pi

The Raspberry Pi 3 Model B+ serves three purposes in our custom UAV. First, it runs the open source ArduSim [21] simulation platform, which controls the UAV at a high-level. This program is capable of coordinating an autonomous flight. Second, it also runs a Python application that processes the camera information (see Figure 5) using the ArUco marker library and sends the resulting information to ArduSim via a TCP connection. Finally, it is used to setup an ad-hoc network in the 5 GHz band. This network can be used to communicate with other UAVs or, in our case, with a ground station (laptop). As an alternative, the Raspberry Pi can be equipped with a 4G LTE dongle so as to operate the UAV from a remote location. However, this option is currently not supported by ArduSim.

4.5. General Components

In addition to the elements referred above, there are mandatory components for each UAV such as:

1. Electronic Speed Controller (ESC): provides power to motors and controls their speed individually with the use of a Pulse Width Modulation (PWM) signal. The signal needed to vary the motor speed is provided by the flight controller.
2. Brushless DC motors to rotate the propellers and thus create thrust.
3. Li-Po battery and power module used to deliver and transmit power to the flight controller, ESCs and all other electrical components.
4. Safety switch which has to be manually pressed by the pilot to ensure no unintended takeoff takes place.
5. An optional buzzer to provide feedback about the current state of the UAV.

5. Experimental Settings

Once we finished the implementation of our solution in ArduSim, we performed validation experiments with the multicopter described in Section 4 in order to assess the effectiveness of our proposed solutions. In the first set of experiments, the UAV was instructed to fly up to an altitude of 20 m, to move toward a specific GPS location and to land automatically (by giving the flight controller full control) once that position was reached. During these experiments, the landing time was recorded, as well as the actual landing position. Those experiments, which do not use our protocol, are used as reference.

In the second set of experiments, the landing accuracy of Algorithm 1 was evaluated. Again, the UAV took off until an altitude of 20 m was reached and then flew towards the target GPS location. The largest available marker (56 × 56 cm) was placed at that location and the UAV used this marker as the initial reference point for landing. When the UAV was able to detect the smaller marker (18 × 18 cm), it used that marker as the reference point instead. For these experiments, the descending speed was defined by lowering the throttle by 10% and the roll and pitch values were set to a value of 5%. After each experiment, the landing time and the distance between the marker and the actual landing position were recorded. We define the landing time as the time interval from the moment when the UAV detected the largest marker until the time when the landing procedure was finished.

In the last set of experiments, the landing time of Algorithm 2 was measured. The markers used were the same. However, in this optimized version, the descending, roll and pitch values were dynamic, as described in Section 3.

6. Results

Without the use of our approach, the UAV was able to land consistently within a time span ranging from 27 to 30 s. Nonetheless, this rapid landing comes at a price. As shown in Figures 7 and 8, the actual landing position varies substantially, ranging from a maximum error of 1.44 m to a minimum error of 0.51 m; the mean value for our experiments was of 0.85 m.

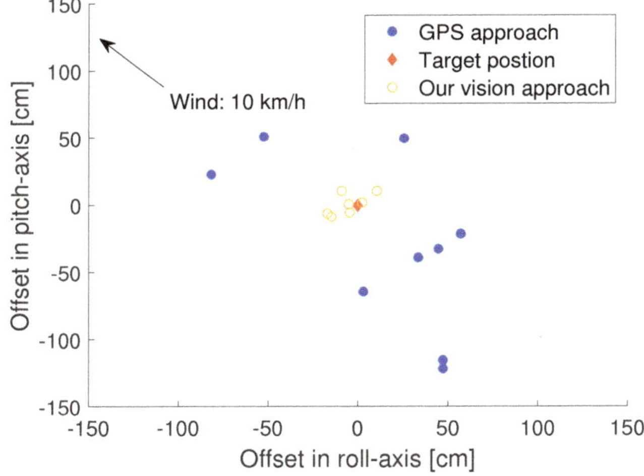

Figure 7. UAV landing position comparison.

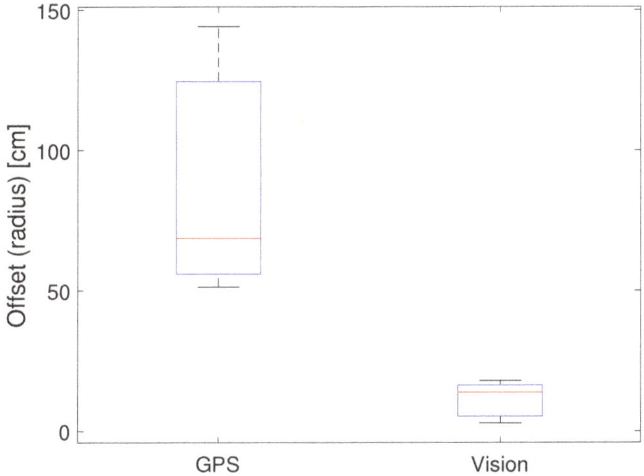

Figure 8. Landing offset Global Positioning System (GPS) vs visual based approached.

Notice that these errors are smaller than expected (1–3 m). This is most likely due to the small travelled distance between the takeoff and landing locations. In fact, in these experiments, the UAV flew for only 14 m and the total flight time was of about 52 s. Longer flights will introduce higher errors, as reported in the literature [22].

As shown in Figure 7, the landing position accuracy increased substantially when Algorithm 1 was adopted. In particular, experiments showed that the error ranged from only 3 to 18 cm, with a mean value of 11 cm (see Figure 8). Overall, this means that the proposed landing approach is able to reduce the landing error by about 96%. However, in three out of ten of the experiments performed, the UAV moved away from the marker due to the effect of wind. Since at these moments the altitude was already quite low, the UAV could no longer detect the marker, causing the mission to be stopped after 30 s. Furthermore, the average landing time was increased to 162 s. This is due to the fact that, during the transition from one marker to another, the algorithm experienced problems at detecting the smaller marker during some time periods, as illustrated in Figure 9a (from second 22 to 35).

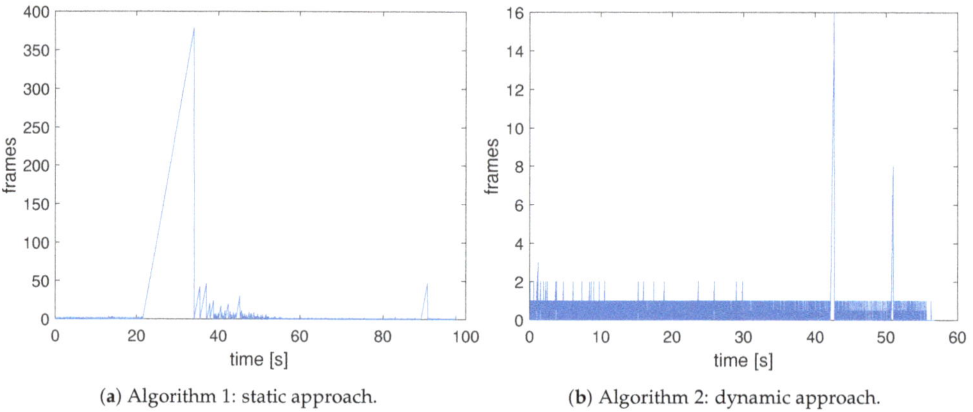

(a) Algorithm 1: static approach. (b) Algorithm 2: dynamic approach.

Figure 9. Number of consecutive dropped camera frames.

Besides this malfunction situations, the UAV showed a smooth landing trajectory (see Figure 10).

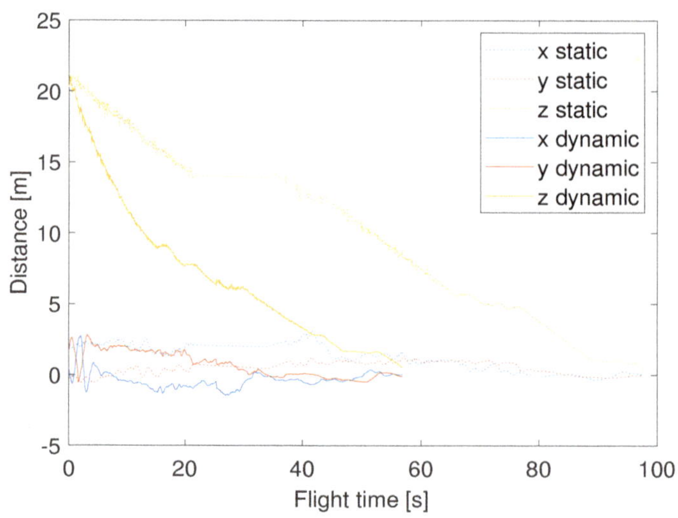

Figure 10. Algortihms: static and dynamic approach.

Notice how the UAV makes more aggressive adjustments in the X axis when the altitude drops below 13 m; this is due to the fact that parameter α becomes smaller, restricting the error range. If the malfunction cases are removed, the average descending speed was of 0.3 m/s, which could be considered too conservative. Furthermore, Figure 10 shows that most of the adjustments are made when the UAV is close to the ground (constant altitude). This can be better observed in Figure 11, where the center of the camera frame and the actual location of the marker in regard to both axes is plotted ($beta_x, beta_y$). It can be seen that the drone only moves when the $beta_x$ or $beta_y$ angle exceeds the value of α. The range of estimated values captured is shown in Figure 12; we can observe that there is a higher variability in the X axis due to wind compensation requirements along that direction during the experiments, something that occurs to a much lower extent for the Y axis.

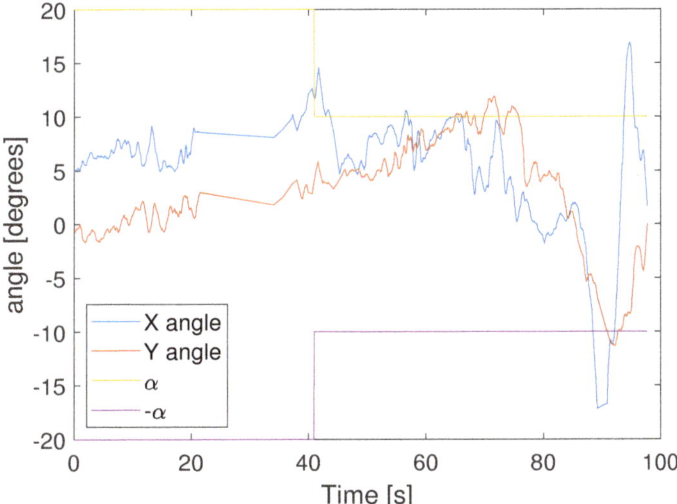

Figure 11. $beta_x$ and $beta_y$ angle variations vs. flight time.

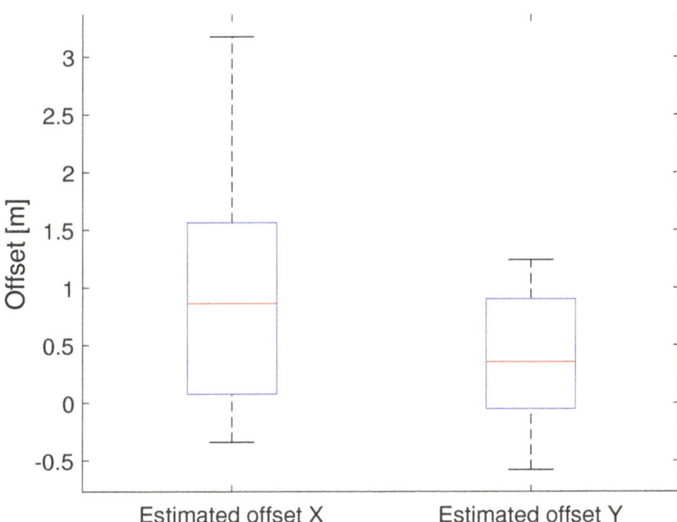

Figure 12. Estimated X and Y variations associated to UAV positions during landing.

The malfunctions of Algorithm 1 are solved in the second version. As shown in Figure 9b, Algorithm 2 does not have any issues when switching between markers. Therefore, the landing time is decreased significantly to an average of only 55 s. Furthermore, as shown in Figure 10, the UAV is descending faster, which also contributes to reduce the landing time. The decrease in landing time did not have any affect on the accuracy of the application. The new recovery mode was also found to be beneficial to ensure that the UAV landed on the marker every time.

To compare the proposed solution to those mentioned in Section 2, Table 3 summarizes the results obtained, highlighting the main differences between them.

Table 3. Comparative table of the different schemes.

Source	Accuracy [m]	Landing Speed [m/s]	Maximum Altitude [m]	Moving Target	Outdoor
Ours/dynamic	0.11	0.3	30	No	Yes
Ours/static	0.11	0.1	20	No	Yes
GPS-based	1-3	0.6	∞	No	Yes
Chen et al. [9]	n/a	0.23	2.5	Yes	No
Araar et al. [13]	0.13	0.06	0.8	Yes	No
Patruno et al. [14]	0.01	n/a	n/a	No	Yes

Finally, an illustrative video (https://youtu.be/NPNi5YC9AeI) has been made available to show how the proposed solution performs in real environments.

7. Conclusions and Future Work

Achieving accurate landing of multirotor UAVs remains a challenging issue, as GPS-based landing procedures are associated with errors of a few meters even under ideal satellite reception conditions, performing worse in many cases. In addition, GPS-assisted landing is not an option for indoor operations. To address this issue, in this work a vision-based landing solution that relies on ArUco markers is presented. These markers allow the UAV to detect the exact landing position from a high altitude (30 m), paving the way for sophisticated applications including automated package retrieval or the landing of large UAV swarms in a very restricted area, among others.

Experimental results using a real UAV have validated the proposed approach, showing that accurate landing (mean error of 0.11m) can be achieved while introducing an additional (but small) time overhead in the landing procedure compared to the standard landing command.

As future work, it is planned to improve the overall efficiency of the protocol. This can be done by improving the flight behaviour. In addition to decreasing the landing time even further, the accuracy can be increased by integrating the UAV's state (acceleration, velocity, etc.) in the control loop. Finally, the algorithm can be further optimized so that the UAV is able to land under less favourable weather conditions.

Author Contributions: C.T.C. conceived the idea; J.W. designed and implemented the protocol with the help of F.F. and T.K.; J.W. analyzed the data and wrote the paper; C.T.C., J.-C.C., F.F., T.K., J.M.M.-B. and P.M. reviewed the paper.

Funding: This work was funded by the "Ministerio de Ciencia, Innovación y Universidades, Programa Estatal de Investigación, Desarrollo e Innovación Orientada a los Retos de la Sociedad, Proyectos I+D+I 2018", Spain, under Grant RTI2018-096384-B-I00.

References

1. Pan, X.; Ma, D.Q.; Jin, L.L.; Jiang, Z.S. Vision-based approach angle and height estimation for UAV landing. In Proceedings of the 1st International Congress on Image and Signal Processing, CISP 2008, Sanya, China, 27–30 May 2008; Volume 3, pp. 801–805. [CrossRef]

2. Tang, D.; Li, F.; Shen, N.; Guo, S. UAV attitude and position estimation for vision-based landing. In Proceedings of the 2011 International Conference on Electronic and Mechanical Engineering and Information Technology, EMEIT 2011, Harbin, China, 12–14 August 2011; Volume 9, pp. 4446–4450. [CrossRef]

3. Gautam, A.; Sujit, P.B.; Saripalli, S. A survey of autonomous landing techniques for UAVs. In Proceedings of the 2014 International Conference on Unmanned Aircraft Systems, ICUAS 2014, Orlando, FL, USA, 27–30 May 2014; pp. 1210–1218. [CrossRef]

4. PX4 Dev Team. Holybro Pixhawk 4 · PX4 v1.9.0 User Guide. 2019. Available online: https://docs.px4.io/v1.9.0/en/flight_controller/pixhawk4.html (accessed on 23 November 2019).

5. Garrido-Jurado, S.; Muñoz-Salinas, R.; Madrid-Cuevas, F.J.; Medina-Carnicer, R. Generation of fiducial marker dictionaries using Mixed Integer Linear Programming. *Pattern Recognit.* **2016**, *51*, 481–491. [CrossRef]

6. Romero-Ramirez, F.J.; Muñoz-Salinas, R.; Medina-Carnicer, R. Speeded up detection of squared fiducial markers. *Image Vis. Comput.* **2018**, *76*, 38–47. [CrossRef]

7. Romero-Ramirez, F.J.; Muñoz-Salinas, R.; Medina-Carnicer, R. ArUco: Augmented reality library based on OpenCV. Available online: https://sourceforge.net/projects/aruco/ (accessed on 7 June 2019).

8. Jin, S.; Zhang, J.; Shen, L.; Li, T. On-board vision autonomous landing techniques for quadrotor: A survey. In Proceedings of the Chinese Control Conference, Chengdu, China, 27–29 July 2016; pp. 10284–10289. [CrossRef]

9. Chen, X.; Phang, S.K.; Shan, M.; Chen, B.M. System integration of a vision-guided UAV for autonomous landing on moving platform. In Proceedings of the IEEE International Conference on Control and Automation, Kathmandu, Nepal, 1–3 June 2016; pp. 761–766. [CrossRef]

10. Nowak, E.; Gupta, K.; Najjaran, H. Development of a Plug-and-Play Infrared Landing System for Multirotor Unmanned Aerial Vehicles. In Proceedings of the 2017 14th Conference on Computer and Robot Vision, Edmonton, AB, Canada, 16–19 May 2017; pp. 256–260. [CrossRef]

11. Shaker, M.; Smith, M.N.R.; Yue, S.; Duckett, T. Vision-Based Landing of a Simulated Unmanned Aerial Vehicle with Fast Reinforcement Learning. In Proceedings of the 2010 International Conference on Emerging Security Technologies, Canterbury, UK, 6–7 September 2010, pp. 183–188. [CrossRef]

12. Lange, S.; Sünderhauf, N.; Protzel, P. Autonomous Landing for a Multirotor UAV Using Vision. In Proceedings of the SIMPAR 2008 International Conference on Simulation, Modeling and Programming for Autonomous Robots, Venice, Italy, 3–7 November 2008; pp. 482–491.

13. Araar, O.; Aouf, N.; Vitanov, I. Vision Based Autonomous Landing of Multirotor UAV on Moving Platform. *J. Intell. Robot. Syst.* **2016**. [CrossRef]

14. Patruno, C.; Nitti, M.; Petitti, A.; Stella, E.; D'Orazio, T. A Vision-Based Approach for Unmanned Aerial Vehicle Landing. *J. Intell. Robot. Syst.* **2019**, *95*, 645–664. [CrossRef]

15. Baca, T.; Stepan, P.; Spurny, V.; Hert, D.; Penicka, R.; Saska, M.; Thomas, J.; Loianno, G.; Kumar, V. Autonomous landing on a moving vehicle with an unmanned aerial vehicle. *J. Field Robot.* **2019**, *36*, 874–891. [CrossRef]

16. De Souza, J.P.C.; Marcato, A.L.M.; de Aguiar, E.P.; Jucá, M.A.; Teixeira, A.M. Autonomous Landing of UAV Based on Artificial Neural Network Supervised by Fuzzy Logic. *J. Control. Autom. Electr. Syst.* **2019**, *30*, 522–531. [CrossRef]

17. Fraczek, P.; Mora, A.; Kryjak, T. Embedded Vision System for Automated Drone Landing Site Detection. In *Computer Vision and Graphics*; Chmielewski, L.J., Kozera, R., Orłowski, A., Wojciechowski, K., Bruckstein, A.M., Petkov, N., Eds.; Springer International Publishing: Cham, Switzerland, 2018; pp. 397–409.

18. Team, A.D. SITL Simulator (Software in the Loop). 2016. Available online: http://ardupilot.org/dev/docs/sitl-simulator-software-in-the-loop.html (accessed on 10 July 2019).

19. Fabra, F.; Calafate, C.T.; Cano, J.C.; Manzoni, P. On the impact of inter-UAV communications interference in the 2.4 GHz band. In Proceedings of the 2017 13th International Wireless Communications and Mobile Computing Conference, Valencia, Spain, 26–30 June 2017; pp. 945–950. [CrossRef]

20. Meier, L.; QGroundControl. MAVLink Micro Air Vehicle Communication Protocol. Available online: http://qgroundcontrol.org/mavlink/start (accessed on 30 January 2019).

21. Fabra, F.; Calafate, C.T.; Cano, J.C.; Manzoni, P. ArduSim: Accurate and real-time multicopter simulation. *Simul. Model. Pract. Theory* **2018**, *87*, 170–190. [CrossRef]
22. Careem, M.A.A.; Gomez, J.; Saha, D.; Dutta, A. HiPER-V: A High Precision Radio Frequency Vehicle for Aerial Measurements. In Proceedings of the 2019 16th Annual IEEE International Conference on Sensing, Communication, and Networking (SECON), Boston, MA, USA, 10–13 June 2019; pp. 1–6. [CrossRef]

 *electronics*

Article

An Efficient and Provably Secure Certificateless Blind Signature Scheme for Flying Ad-Hoc Network Based on Multi-Access Edge Computing

Muhammad Asghar Khan [1], **Ijaz Mansoor Qureshi** [2], **Insaf Ullah** [3], **Suleman Khan** [2], **Fahimullah Khanzada** [4] **and Fazal Noor** [5,*]

[1]   Department of Electronic Engineering, ISRA University, Islamabad 44000, Pakistan; khayyam2302@gmail.com
[2]   Department of Electrical Engineering, AIR University, Islamabad 44000, Pakistan; imqureshi@mail.au.edu.pk (I.M.Q.); 171518@students.au.edu.pk (S.K.)
[3]   Department of Computer Sciences, Hamdard University, Islamabad 44000, Pakistan; insafktk@gmail.com
[4]   Descon Engineering Limited, Lahore 54000, Pakistan; fahimullahk@gmail.com
[5]   Department of Computer and Information Systems, Islamic University of Madinah, Madinah 400411, Saudi Arabia
*   Correspondence: mfnoor@gmail.com; Tel.: +966-551497218

Received: 31 October 2019; Accepted: 22 December 2019; Published: 26 December 2019

Abstract: Unmanned aerial vehicles (UAVs), when interconnected in a multi-hop ad-hoc fashion, or as a flying ad-hoc network (FANET), can efficiently accomplish mission-critical tasks. However, UAVs usually suffer from the issues of shorter lifespan and limited computational resources. Therefore, the existing security approaches, being fragile, are not capable of countering the attacks, whether known or unknown. Such a security lapse can result in a debilitated FANET system. In order to cope up with such attacks, various efficient signature schemes have been proposed. Unfortunately, none of the solutions work effectively because of incurred computational and communication costs. We aimed to resolve such issues by proposing a blind signature scheme in a certificateless setting. The scheme does not require public-key certificates, nor does it suffer from the key escrow problem. Moreover, the data that are aggregated from the platform that monitors the UAVs might be too huge to be processed by the same UAVs engaged in the monitoring task. Due to being latency-sensitive, it demands high computational capability. Luckily, the envisioned fifth generation (5G) mobile communication introduces multi-access edge computing (MEC) in its architecture. MEC, when incorporated in a UAV environment, in our proposed model, divides the workload between UAVs and the on-board microcomputer. Thus, our proposed model extends FANET to the 5G mobile network and enables a secure communication between UAVs and the base station (BS).

Keywords: blind signature; security; MEC; UAVs; FANET; 5G; IoT

---

1. Introduction

During the last couple of years, the exponential advancement in the manufacturing of small unmanned aerial vehicles (UAVs) has led to a new clan of networks, referred to as flying ad-hoc network (FANET). The prominent features of agility, low-cost, and easy deployment, among others, are paving ways for FANET to offer successful solutions for diverse military and civilian application. In case of a disastrous situation, FANET can offer a cost-effective solution for real-time data communication as compared to its predecessors, these being, mobile ad-hoc networks (MANETs) and vehicular ad-hoc networks (VANETs) [1]. Not only does FANET have the capability of collecting and sharing the aggregated data amongst the UAVs, it can also send it to the base station (BS). Additionally, if some of the UAVs are detached during the mission, irrespective of any reason, they still have the facility to remain associated to the network with the support of other UAVs due to an ad-hoc network between the UAVs. Furthermore, the inherent multi-hop networking schema counters the obstacles of short-range communication and limited guidance that normally arise in a stand-alone UAV system [2]. Nevertheless, such exclusive attributes make FANET a suitable solution for various applications. The small UAVs have restricted capabilities in terms of power, sensing, communication, and computation. This renders the small UAVs as luring targets to different kinds of known and unknown cyber-attacks. Generally, in a FANET environment, multiple UAVs are integrated into a team that cooperates with each other to accomplish critical tasks [3]. Hence, when a self-governing UAV desires to perform a certain task, it receives the command containing relevant task-specific information such as time, target location, and actions, among others. Then, it either flies autonomously to the target position in the assigned time, or it may cruise in the air while waiting for commands, thus reducing the response time and accomplishing the results proficiently.

The ground station interconnects with UAVs over an unauthenticated and unencrypted channel. Therefore, anyone with a suitable transmitter can link with the UAV and insert commands into an ongoing session, and thus can easily interpret any UAV. Thus, it is important for a UAV to ascertain the origination of a command. Normally, digital signatures are used to ascertain the source of command. He et al. [4] described the overall process as follows:

(1) A command center initiates command and computes the corresponding digital signature.
(2) The corresponding command and signature are then forwarded to the UAV by the command center.
(3) The UAV, upon receiving the command and signature, attempts to verify the signature.

- If the signature is valid, the UAV deems it to be issued by the command center and proceeds with executing the command.
- Otherwise, the command is considered counterfeit and, thus, the UAV does not execute it.

However, due to its intrinsic complexities and security requirements, the mutual digital signature scheme is not appropriate for an UAV-based network. Additionally, the average speed of a typical UAV can lie in the range of 30–460 km/h in a three-dimensional (3D) setting [5]. Moreover, the topology of the particular network varies rapidly, which necessitates the need of ascertaining the validity of a command in the shortest time. Therefore, it is essential for the UAV to validate the signature in a timely manner, especially for location-based services. For example, the user or ground station (GS) pledges a command and the corresponding signature to the UAV; however, the concerned UAV can only verify the signature. Even in the worst case, if an intruder eavesdropped on the command and corresponding signature, they cannot authenticate the signature and authorize the task to be accomplished next. In addition, frequent changes in topology also increase the latency and communication cost. In order to accommodate the key escrow problem, a certificateless signature scheme is required. In a certificateless cryptosystem, a participant private key is composed of two parts: the partial private key and a secret value. The trusted third-party key generation center (KGC) generates the partial private key, whereas the secret value is affirmed by the participant. Similarly, a participant's public key also consists of two parts, these being the participant's identity information and the public key conforming to the secret value. Therefore, the cost of public key management is significantly reduced due to the fact that the public key does not require any certificate. Furthermore, it does not suffer from the key escrow problem because the KGC has no information about the participant secret value.

Typically, small UAVs have batteries that last for merely 20 to 30 min [6]. Therefore, it is of utmost importance to manage the battery resources efficiently. This prolongs the network lifetime especially for large-scale deployments of UAVs. Thus, it is harder for the UAVs to complete these resource-hungry applications in a timely fashion. Furthermore, FANET can be deployed in remote locations to assist the Internet of Things (IoT) devices for collecting large volumes of data. Fortunately, these impediments can be mitigated by employing multi-access edge computing (MEC) technology. The MEC shifts the job performed by the commanding UAV to the edge of the network, which is closer to UAV, thus reducing the propagation delay. The MEC thus paves way for a diverse set of applications that, explicitly, demand a real-time response. The heterogeneous radio access network of a ground-based network is composed of macro cells and small cells. The network assists the mobile phones, driver-less cars, and IoT gadgets, among others, in performing the required operations. Therefore, as a direct consequence, a multitude of emerging technologies can synergize with the 5G (fifth generation) wireless networks. A symbiotic relation can be visualized between the UAVs, engaged in scheduling the computing tasks, and the onboard microprocessor, dedicated to executing the particular operations. Furthermore, the usable data can be stored temporarily for retrieval by either the UAVs or the ground devices, while, concurrently, the drone-cells transmit the data.

Normally, the security and efficiency of the aforementioned signature scheme is based on some computationally hard problems, for example, Rivest–Shamir–Adleman (RSA), bilinear pairing, and elliptic curve cryptosystems (ECC). RSA offers a solution based on large factorization [7,8], which utilizes a 1024 bit large key [9]. However, due to the restricted on-board processing capabilities on UAVs, the solution is not appropriate for the resource-constrained FANET system. In addition, bilinear pairing, which suffers from high pairing and map-to-point function computations, is 14.90 times worse than RSA [10]. Therefore, in order to counter the shortcomings of RSA and bilinear pairing, a new category of cryptography, elliptic-curve cryptography (ECC) was introduced [11]. ECC is characterized by a smaller parameter size and involves miniaturized versions of public key, private key, identity, and certificate size, among other factors. Moreover, unlike bilinear pairing and RSA, the security hardiness and efficiency of the scheme is based on 160 bit small key, which is still not suitable for resource-hungry devices [12]. Thus, a new type called hyperelliptic-curve cryptography (HECC) was proposed [13]. The hyperelliptic curve uses an 80 bit key, identity, and certificate and offers security to the degree comparable to that of elliptic curve, bilinear pairing, and RSA [14,15]. It is, hence, a far better choice for energy-constrained devices.

1.1. Authors' Motivations and Contributions

A comprehensive literature review of the existing blind signature schemes was carried out. It was found that these schemes are based on hard problems, that is, elliptic curve, bilinear pairing, and modular exponential, and thus suffer from high computational and communication costs. Hence, the existing schemes are not compatible with small devices, that is, UAVs that have limited computational power. Moreover, these schemes are not validated using formal security validation tools such as automated validation of internet security protocols and applications (AVISPA) or Scyther, among others, which can, somehow, guarantee security. There is a critical need to harness the state-of-the-art certificateless blind signature scheme so as to engineer a viable cryptographic solution for FANET that poses less danger to the battery lifetimes of resource-constrained UAVs.

The authors, motivated by the aforementioned objectives, to name a few, propose a new scheme, called provably verified certificateless blind signature (CL-BS) scheme for FANET. The proposed scheme is based on hyperelliptic curve, which is an advanced version of the elliptic curve. It provides the same level of security as elliptic curves, bilinear pairing, and modular exponential with smaller key size. Some of the salient features signifying contributions of our research work, in this paper, are as follows:

- We introduce a novel architecture for flying ad-hoc network (FANET) constituted by UAVs with a multi-access edge computing (MEC) facility that leverages the 5G wireless technology.
- We propose an efficient and provably secure certificateless signature (CL-BS) scheme for the same architecture using the concept of hyperelliptic curve for operating in resource-constrained environments.
- The proposed scheme is shown to be resistant against various attacks through formal as well as informal security analysis using the widely-accepted automated validation for internet security validation and application (AVISPA) tool.

- The proposed scheme is also compared with existing counterparts and it is shown that our approach provides better efficiency in terms of computational and communication costs.

1.2. Structure of the Paper

The remainder of the paper is structured as follows: Section 2 contains a brief about the related work; Section 3 presents the foundational concepts; Section 4 presents the proposed architecture and construction of proposed scheme (i.e., CL-BS); Section 5 holds implementation of the proposed scheme in FANET; Section 6 outlines the AVISPA tool component of our proposed scheme for formal security verification as well as informal security analysis; Section 7 compares the proposed scheme with the existing schemes; and in the end, Section 8 succinctly culminates the manuscript by concluding the work.

2. Related Work

2.1. Flying Ad-Hoc Network

In flying ad-hoc network, the security and privacy are important because UAVs are always unattended. The primary security mechanisms for FANET emphasize authenticity, confidentiality and integrity of data via cryptography. A well-designed data protection mechanism can significantly reduce the probability of the data becoming compromised, irrespective of the malicious technique involved. There are a few studies dedicated to investigating the data protection issues for UAV networks. Won et al. [16], proposed a suite of cryptographic protocols for drones and smart objects. The protocols deal with three communication scenarios: one to-one, one-to-many, and many-to-one. In the first scenario, that is, one-to-one, the efficient encapsulation mechanism, a certificateless signcryption tag key, backs the authenticated key agreement in addition to offering non-repudiation and user revocation. The one-to-many scenario involves a certificateless multi-recipient encryption scheme, which allows a UAV to transmit privacy-intensive data to multiple smart objects. Lastly, UAVs are able to collect data from multiple smart objects in the "many-to-one" communication scenario. The protocol, however, finds it difficult to transmit a multitude of encrypted messages and at the same time assure privacy of the end devices. Such novel cryptographic mechanisms are efficient and secure. However, they are supposed to be used in group communication where nodes are of equal computational capability. A novel approach to mitigate the broadcast storm problem during the interest's dissemination is proposed by Barka et al. [17]. The approach is based on a trust-aware monitoring communication architecture for flying named data networking. It makes use of the inter-UAV communication for checking the data authenticity on a particular UAV without disturbing the desired level of security. However, data privacy and caching policies are not taken into consideration in the proposed scheme.

In order to resist the physical capturing of drones with minimum exposure of confidential data, Bae et al. [18] proposed a saveless-based key management and delegation system for a multi-drone environment. Nevertheless, the proposed scheme is not compatible with devices such as UAVs equipped with limited on-board energy that hinder the potency of finding a proper key renewal period with low computation cost and more security guarantee. Seo et al. [19] proposed a pairing-free approach for drone-based surveillance applications. However, this approach faces the user revocation problem in the case of a physical attack. In such a case, the intruders can access not only current but future information of the drones. In order to cater the forward secrecy problem in drones, Liu et al. [20] proposed two construction schemes that achieve better performance in terms of the computational cost required by the recipient. However, the approach uses the elliptic curves and, thus, it suffers from high computational cost. Moreover, the proposed scheme is not validated through formal security analysis. In 2018, Reddy et al. [21] presented a pairing-free key insulated signature scheme in the identity-based setting for improving the computational and communication efficiency. Later, in 2019, Xiong et al. [22] also proposed a pairing-free and provably secure certificateless parallel key-insulated signature (CL-PKIS) scheme in order to secure the communication in the IIoT setting. However, because the scheme involves the concept of elliptic curve, it is not free from the issue of high computational cost. Moreover, the proposed schemes are not validated through formal security analysis.

2.2. Multi-Access Edge Computing

It is mandatory for a FANET system to diminish latency to the maximum possible extent. MEC can solve the problem of latency resulting from long communication distance. So far, studies have been conducted to examine the usage of edge computing for UAVs [23–25]. However, the studies do not discuss the topic of communication link quality. ETSI proposed a reference architecture of MEC [26]. Primarily, the MEC reference architecture is composed of user equipment, mobile edge applications, and networks. The network is classified into either of the following three levels: system level, host level, and network level. The reference points and functional elements of the reference architecture are depicted by the reference architecture. Garg et al. aimed to answer the surveillance-related concerns by proposing a data-driven transportation optimization model [27]. The model comprises UAV, dispatcher, aggregator, and edge devices. Each of the constituents undertake the designated tasks as follows: the UAV captures and validates the date; the dispatcher, in addition to validating the tasks, schedules the processing tasks in the edge computing devices; the aggregator assures a secure transmission of data; and the edge devices analyze the data. A hierarchical MEC architecture has been proposed by Lee et al. [28]. It involves utilizing the resources of the MEC server for providing services customized on the basis of content type and the computing demand. After exploring the major causes of communication and computational latencies, Intharawijitr et al. proposed a mathematical model. The model is used to estimate the computing latency in an edge node selected on the basis of either of the three policies [29]. A game theoretic model is proposed by Messous et al. in which the UAVs, as game players, strategize to achieve the optimal tradeoff between energy overhead and the execution delay. As a result, the UAVs do not stay overburdened anymore [30]. Ansari et al. addressed the issue of end-to-end delay between the proxy virtual machine and the device. They claim to resolve the problem by suggesting two dynamic proxy virtual machine migration methods, which is corroborated by simulation results [31]. Zhang et al. attempted to resolve the issue of increased energy consumption and longer execution time by proposing a mobility-aware hierarchical MEC framework [32]. The proposed solution involves the MEC servers and, for sharing the computing tasks, a backup computing server. An incentive-based optimal computational offloading scheme was developed. The objective of quick-response and energy conservation was achieved to a significant extent.

The methodology proposed by Christian et al. [33] increases the system reliability and reduces the end-to-end source-actuator latency. Their work intends to broaden the 5G network edge by making the FANET UAVs fly close to the monitoring layer. The UAVs are accoutered with MEC facilities while carrying out the processing tasks and they follow a policy for mutual help for improved performance. However, the work fails to address the issue of limited battery duration of MEC UAVs.

2.3. Related Certificateless Blind Signature Schemes

As early as 1983, Chaum presented a blind signature scheme that significantly reduces the probability of detectability [34]. The scheme, for the case of transmitting a message, revolves around two major players: signer, the entity that computes the signature, and provider, the part tasked to blind the message. The signer transmits the computed signature to the provider, who deciphers and retrieves the original signature. Owing to its versatility, the scheme can, in e-commerce settings, help establish a forgery-resistant payment system. In 1996, Mambo et al. proposed a proxy signature scheme in which the original signer can delegate the task of issuing signature to a proxy security and communication network 5 [35]. Tan et al. applied the concepts of discrete logarithm and elliptic curve discrete logarithm to suggest two proxy blind signature schemes [36]. Each of the schemes offers the security threshold promised by the proxy and blind signature schemes. Tan proposed another proxy blind signature scheme as well [37]. The comparatively efficient scheme is based on identity and is pairing-free. It proves to be secure in random oracle model. A proxy partially blind signature scheme has been proposed by Yang et al. The scheme can revoke the proxy privileges and is characterized by security features [38]. Verma et al. proposed a proxy blind signature scheme [39]. The scheme exhibits message recovery and caters to the requirements of low-bandwidth because it abbreviates the size of message signature. The efficient identity-based proxy blind signature scheme proposed by Zhu et al. can even overcome a quantum computer attack [40]. A designated verifier signature scheme is proposed by Jakobsson et al. [41]. Dai et al. further advanced the concept of designated receiver proxy signature scheme by presenting a combo of a designated verifier signature scheme and a proxy signature scheme [42]. In the schema, a proxy signer is delegated the authority to sign, and then authenticate, in lieu of the original signer.

A short-designated verifier proxy signature (DVPS) scheme is proposed by Huang et al. [43]. The scheme is characterized by signatures of comparatively shorter length and, thus, caters to the applications requiring low bandwidth. Shim furthered the idea by presenting a short DVPS scheme based on BLS signature. The scheme proves to be superior when tested using the random oracle model [44]. Islam et al. considered the concept of bilinear pairing to propose an efficient identity-based DVPS scheme [45]. The scheme assigns private keys to the involved entities generated from a private key generator (PKG). Hu et al. proposed two DVPS schemes: weak DVPS and strong DVPS [46]. Although the weak DVPS scheme is not able to compute a DVPS, the strong DVPS scheme can do so. The random oracle model is used to prove the effectiveness of the proposed scheme. It has been demonstrated on multiple occasions that the blind signature scheme offers forgery-proof operations when applied in sensitive applications such as e-voting and e-cash, among other applications [47,48]. However, anonymity and intractability of the voter and the unforgeability of the electronic vote are the main security concerns. In addition, the proposed schemes are based on bilinear pairing and elliptic curves, both of whom are costly operations in cryptography. Chin et al. [49] presented a certificateless blind signature scheme based on bilinear pairing. Likewise, the proposed scheme is based on bilinear pairing. Furthermore, the security analysis is done through random oracle model and has not been authenticated using any tool.

3. Preliminaries

A brief overview of some of the foundational concepts, along with their formal definitions, is presented in this section.

3.1. Hyperelliptic Curve Cryptosystems (HECC)

Hyperelliptic curves can be viewed as generalizations of ECC (elliptic curve cryptosystems), introduced by Koblitz [50]. A hyperelliptic curve [51] is denoted over curves, whose genus is greater than 1, as shown in Figure 1. Similarly, the curves with genus 1 are generally known as elliptic curves. The group order of the field F_q for genus 1, 160 bit long operands are required, that is, we need at least $g.\log_2(q) \approx 2^{160}$, where g is the genus of curve over F_q that is a set of finite fields of order q. Likewise, for curves with genus 2, 80 bit long operands, and, for curves with genus 3, 54 bit long operands are needed [52].

A hyper elliptic curve C of genus greater than 1 over F is a set of solutions $(x, y) \in F \times F$ to the following equation:

$$C: y^2 + h(x)y = f(x). \tag{1}$$

A divisor D is a finite formal sum of points on hyper elliptic curve and represented as

$$D = \sum_{P_i \in C} m_i p_i , \; m_i \in Z. \tag{2}$$

The two divisors can be added as follows:

$$\sum_{P_i \in C} m_i p_i + \sum_{P_i \in C} n_i p_i = \sum_{P_i \in C} (m_i + n_i) p_i. \tag{3}$$

Each element of the Jacobian can be represented in the semi-reduced divisor form [53]:

$$D = \sum_i m_i p_i - \left(\sum_i m_i \right), \ \forall mi \geq 0. \tag{4}$$

If the divisor is subjected to the additional constraint, that is, $r \leq g$, such a divisor is defined as a reduced divisor. Additionally, in [50], the author shows that the divisors of the Jacobian can be denoted as a pair of polynomials $a(x)$ and $b(x)$ with following degrees: $b(x) \leq \deg a(x) \leq g$, where the coefficients of $a(x)$ and $b(x)$ are elements of F and $a(x)$ divided by $y^2 + h(x)y - f(x)$.

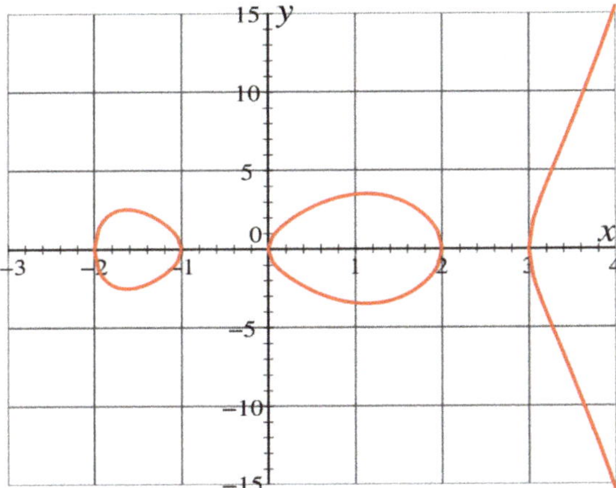

Figure 1. Hyper elliptic curve of genus 2, that is, g = 2.

3.2. Threat Model

The widely used Dolev–Yao (DY) threat model [54] is used in the proposed scheme. According to the DY model, an insecure public channel (open channel) is used for communication between any two parties and the end-point entities have an untrustworthy nature. Therefore, the system is prone to eavesdropping of exchanged messages and deletion/modification attempts by the attacker. Moreover, as the UAVs may roam around in unattended hostile areas, there exists the probability of them getting physically captured. This may lead to leakage of precious data from a UAV's memory. The KGC, on the other hand, is a fully trusted entity.

4. Proposed Architecture

The proposed architecture of flying ad-hoc network based on multi-access edge computing is illustrated in Figure 2. The application scenario considered is the surveillance of a specific area, which may collect data, that is, video streaming and images. We consider two representative classes of UAVs: monitoring UAV (M-UAV) and raspberry pi-based multi-access edge computing UAV (RMEC-UAV). M-UAV perform data acquisition and monitoring only from the assigned zone. In our proposed architecture, the set of M-UAVs are assigned to one RMEC-UAV that is used to reduce the power consumption while executing the security mechanism (i.e., sign, verify). The set of M-UAVs allocated to RMEC-UAV is essentially subjected to the load produced by M-UAV. RMEC-UAV collects data from M-UAVs and forwards this to the base station. RMEC-UAV can also connect with the IoT devices and collect data from them. Prior to transmitting, the RMEC-UAV validates the authenticity of the M-UAVs. Upon successful validation, the RMEC-UAV forwards the data to the BS. The RMEC-UAV transmits not only the IoT data but also the flight information and the information about the role of each M-UAV.

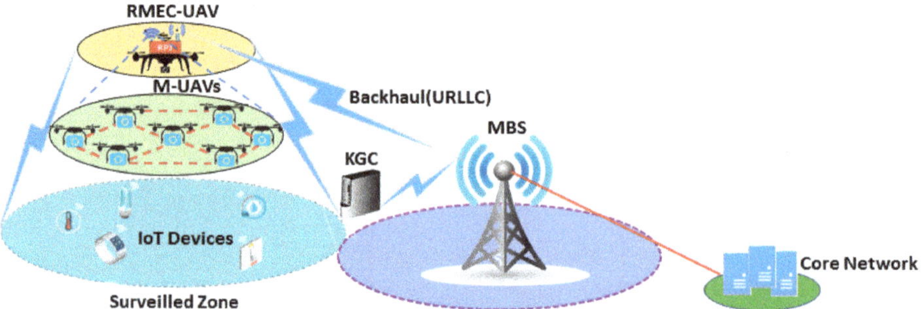

Figure 2. Proposed architecture of flying ad-hoc network with 5G and multi-access edge computing (MEC) facilities. RMEC-UAV: raspberry pi-based multi-access edge computing unmanned aerial vehicle, M-UAV: monitoring UAV, IoT: Internet of Things, URLLC: ultra-reliable low latency communications, KGC: key generation center.

Raspberry PI (RPI) board was considered for RMEC-UAV. Even though there are other substitutions for RPI, with sophisticated hardware configurations, such as LattePanda 4G/64 GB, Qualcomm Dragon board, ODROID-XU4, and ASUS Tinker Board, among others, RPI is nonetheless considered to be the most cost-effective and energy-efficient option. Other alluring features of RPI 4 that further defend its selection are the built-in wireless networking support, that is, Wi-Fi (dual-band 802.11 b/g/n/ac) and Bluetooth 5.0 BLE. RPI 4 is equipped with a 1.5 GHz 64-bit quad core ARM Cortex-A72 processor. The 5G and 802.11 ac wireless modules are enabled on RMEC-UAV in order to link it with the BS/IoT devices and hence provide a hotspot service over the M-UAVs. Fifth generation (5G) is further classified into enhanced mobile broadband (eMBB), massive machine-type communications (mMTC), and ultra-reliable low latency communications (URLLC) by the International Telecommunication Union (ITU) in order to fulfill the requirements of diverse industrial and market demands. However, we have considered URLLC in the proposed architecture, as of it offers very high mobility, which further defends its selection for UAV-based operations [55].

The images transmitted by the monitoring UAVs, ground cameras, and the sensors, among other sources, are all received by the RMEC-UAV on-board microcontroller. The microcontroller, then, generates the tasks that will be processed by the local microcomputer, or the decision support engine (DSE). The human operator receives a decreased share of the data flow so as to decide quickly. In case the human decisions are not timely, the predictive and interpolative/extrapolative modules mounted on the RMEC-UAV DSE step in. The probability of response-delays resulting from the queues of to-be-processed jobs can never be ignored. To compensate for such time lapse and to enhance reliability, the RMEC-UAVs synergize with each other. Further, each of the M-UAVs, after being equipped with the essential gadgets, these being cameras, IMU, sensors, and a GPS unit, among others, can be accustomed to different application scenarios.

The proposed architecture can be divided into the following three main layers:

- **Layer 1** consists of the ground-level IoT devices that are devoted to different tasks as per application scenario. The ground-level IoT devices are connected with the RMEC-UAV and BS via URLLC, a 5G wireless link. Furthermore, the macro base station (MBS) are typically linked with the core network via wires that have huge bandwidth.

- **Layer 2** comprises a team of M-UAVs equipped with the essential gadgets, these being cameras, IMU, sensors, and a GPS unit, among others, for monitoring the assigned zone. Moreover, M-UAVs are connected with each other using Bluetooth 5 (2.4 GHz) link and with the RMEC-UAV with 802.11 ac (5 GHz) Wi-Fi link.
- **Layer 3** is composed of RMEC-UAV that is used to collect data from M-UAVs and forwards it to the base station. RMEC-UAV can also connect with the ground-level IoT devices and collect data from them.

Construction of the Proposed Scheme

The proposed scheme includes the following four entities: KGC, blind signer, requester, and verifier. Further, it involves the following six sub-algorithms for producing the certificateless blind signature: setup, partial private key setting (PPKS), secret value setting (SVS), private key setting (PKS), public key setting (PBKS), blind signature, and verification. In Table 1, we provide an explanation about the notations used in the proposed algorithm. Therefore, for representing the whole process of certificateless blind signature, we aimed to provide the simplest explanation by using the following steps:

1. Setup: In this sub-algorithm, the KGC selects the following parameters:

   - A hyper elliptic curve (C);
   - A divisor ($\mathcal{D}$), where D is the divisor in C;
   - The hash function (h);
   - Select ∂ from $\{1, 2, \ldots, n-1\}$ and the size of $n = 2^{80}$.

   After the above process, the KGC determines the master public key using $Y = \partial.\mathcal{D}$. Then, it publishes the set of selected parameters, $\{C, \mathcal{D}, Y, n = 2^{80}\}$.

2. Partial private key setting (PPKS): In order to set the partial private key for the participating users (verifier and signer) with identity $\mathcal{JD}_\sqcap$, the KGC performs the following sub-steps:

   - It selects X_u from $\{1, 2, \ldots, n-1\}$;
   - It computes $\alpha_u = X_u. \mathcal{D}$ and $\beta_u = X_u + \partial. \alpha_u$;
   - It computes $\delta_u = \beta_u. \mathcal{D}$;
   - It sends (β_u, δ_u) to the users (verifier and signer) with identity $\mathcal{JD}_\sqcap$.

   The users can verify the pair (β_u, δ_u) such as: $\beta_u. \mathcal{D} = \alpha_u + \alpha_u.Y = \alpha_u + \alpha_u.Y = X.\mathcal{D} + \alpha_u (\partial.\mathcal{D}) = \mathcal{D}(X + \alpha_u. \partial) = \mathcal{D}(\beta_u) = \beta_u.\mathcal{D}$.

3. Secret value setting (SVS): The user (verifier and signer) with identity $\mathcal{JD}_\sqcap$ selects Q_u from $\{1, 2, \ldots, n-1\}$ and keeps it as his secret value.
4. Private key setting (PKS): The user (verifier and signer) set the private key as $\sigma_u = \langle Q_u, \beta_u \rangle$.
5. Public key setting (PBKS): The user (verifier and signer), with identity $\mathcal{JD}_\sqcap$, compute $\chi_\sqcap = \sigma_u.\mathcal{D}$.

   The user sets his/her public key as $\mathcal{B}_u = \langle \chi_\sqcap, \delta_u \rangle$.

   - Blind signature: In this part, the blind signer first selects ω from $\{1, 2, \ldots, n-1\}$, computes $\Delta_1 = \omega/Q_s$, $\Delta_2 = \beta_s/\omega$, and then sends it (Δ_1, Δ_2) to the requester. Further, the requester proceeds as follows:
   - It selects (τ, φ) from $\{1, 2, \ldots, n-1\}$;
   - It computes $\mathcal{E} = h(\mathcal{I}, \Delta_1, \Delta_2, \sigma_r)$ and $Z = \mathcal{E} + \varphi$;
   - It sends Z to the blind signer. The blind signer generates the partial blind signature $\mathcal{S}^* = Q_s - Z.\beta_s$ and sends it to the requester;
   - The requester, then, generates the hash value as $V = (\mathcal{I}, Ns)$ and full blind signature, using $\mathcal{S}^{**} = \mathcal{S}^* - \tau$, and transfers it $(\mathcal{S}^{**}, V)$ to the verifier.

6. Verification: The verifier can verify the blind signature if either of the following equalities are satisfied: $V^* = (\updownarrow, Ns) = V = (\updownarrow, Ns)$ or $V^* = V$.

Table 1. Notations used in proposed algorithm.

S.NO	Symbol	Definition
1	C	Means hyperelliptic curve with genus 2, which is the generalized form elliptic curve requiring 80 bit key
3	$\mathcal{D}$	A divisor, which is a finite formal sum of points on hyperelliptic curve
4	∂	Master secret key which is generated by KGC for producing partial private key
5	Y	The master public key of KGC
6	N	Randomly generated number and the size of N as $n = 2^{80}$
7	$\hbar$	One-way hash function, which means that it has the property of irreversibility
8	$\sigma_r = <Q_r, \beta_r>$	Private key of requester
9	$\sigma_s = <Q_s, \beta_s>$	Private key of signer
10	Ns	A fresh nonce that is used for anti-replay attack
11	$\mathcal{ID}_\sqcap$	Identities for sender and receiver
12	$\updownarrow$	Plain-text (message)

5. Implementation of Proposed Scheme in FANET

We divided this process in two sub-phases that are (1) initialization and registration, and (2) signing and verifying Phase, which are illustrated in Figures 3 and 4, respectively.

5.1. Initialization and Registration

In this sub algorithm, the KGC selects a hyper elliptic curve (C), a divisor ($\mathcal{D}$), the hash function ($\hbar$), and then computes ∂ from $\{1, 2, \dots, n - 1\}$ and the size of $n = 2^{80}$.

After the above process, the KGC determines the master public key using $Y = \partial.\mathcal{D}$. Then, it publishes the set of selected parameters, $\{C, \mathcal{D}, Y, n = 2^{80}\}$.

- **Partial Private Key Generation for RMEC-UAV:** In order to set the partial private key for *RMEC-UAV* with identity $\mathcal{ID}_{rv}$, the KGC performs the following sub steps:

  i. It selects X_{rv} from $\{1, 2, \dots, n - 1\}$;
  ii. It computes $\alpha_{rv} = X_{rv}.\mathcal{D}$ and $\beta_{rv} = X_{rv} + \partial. \alpha_{rv}$;
  iii. It computes $\delta_{rv} = \beta_{rv}.\mathcal{D}$;
  iv. It sends $(\beta_{rv}, \delta_{rv})$ to the *RMEC-UAV*.

The *RMEC-UAV* can verify the pair $(\beta_{rv}, \delta_{rv})$, such as $\beta_{rv}.\mathcal{D} = \alpha_{rv} + \alpha_{rv}.Y = \alpha_{rv} + \alpha_{rv}.Y = X.\mathcal{D} + \alpha_{rv}.(\partial.\mathcal{D}) = \mathcal{D}(X + \alpha_{rv}.\partial) = \mathcal{D}(\beta_{rv}) = \beta_{rv}.\mathcal{D}$.

- **Secret Value Setting for RMEC-UAV:** The *RMEC-UAV* selects Q_{rv} from $\{1, 2, \dots, n - 1\}$ and keeps it as their secret value.
- **Private Key Setting for RMEC-UAV:** The *RMEC-UAV* sets the private key as $\sigma_{rv} = <Q_{rv}, \beta_{rv}>$.
- **Public Key Generation for RMEC-UAV:** The RMEC-UAV computes $\chi_{rv} = \sigma_{rv}.\mathcal{D}$ and sets their public key as $\mathcal{B}_{rv} = <\chi_{rv}, \delta_{rv}>$.
- **Partial Private Key Setting for BS/IoT:** In order to set the partial private key for *BS/IoT* with identity $\mathcal{ID}_{BI}$, the KGC performs the following sub steps:

  i. It selects X_{BI} from $\{1, 2, \dots, n - 1\}$;
  ii. It computes $\alpha_{BI} = X_{BI}.\mathcal{D}$ and $\beta_{BI} = X_{BI} + \partial. \alpha_{BI}$;

iii.   It computes $\delta_{\mathrm{BI}} = \beta_{\mathrm{BI}}.\,\mathcal{D}$;

iv.   It sends $(\beta_{\mathrm{BI}}, \delta_{\mathrm{BI}})$ to *BS/IoT*.

The *RMEC-UAV* can verify the pair $(\beta_{\mathrm{BI}}, \delta_{\mathrm{BI}})$, such as $\beta_{\mathrm{BI}}.\mathcal{D} = \alpha_{\mathrm{BI}} + \alpha_{\mathrm{BI}}.Y = \alpha_{\mathrm{BI}} + \alpha_{\mathrm{BI}}.Y = X.\mathcal{D} + \alpha_{\mathrm{BI}}.(\partial.\mathcal{D}) = \mathcal{D}\,(X_{\mathrm{BI}} + \alpha_{\mathrm{BI}}.\ \partial) = \mathcal{D}\,(\beta_{\mathrm{BI}}) = \beta_{\mathrm{BI}}.\mathcal{D}$.

- ***Secret Value Setting for BS/IoT:*** The *BS/IoT* selects Q_{BI} from $\{1, 2, \dots, n-1\}$ and keeps it as their secret value.
- ***Private Key Setting for BS/IoT:*** The *BS/IoT* sets the private key as $\sigma_{\mathrm{BI}} = <Q_{\mathrm{BI}}, \beta_{\mathrm{BI}}>$.
- ***Public Key Setting for BS/IoT:*** The *BS/IoT* computes $\chi_{\mathrm{BI}} = \sigma_{\mathrm{BI}}.\mathcal{D}$ and sets their public key as $\mathcal{B}_{\mathrm{B}} = <\chi_{\mathrm{BI}}, \delta_{\mathrm{BI}}>$.

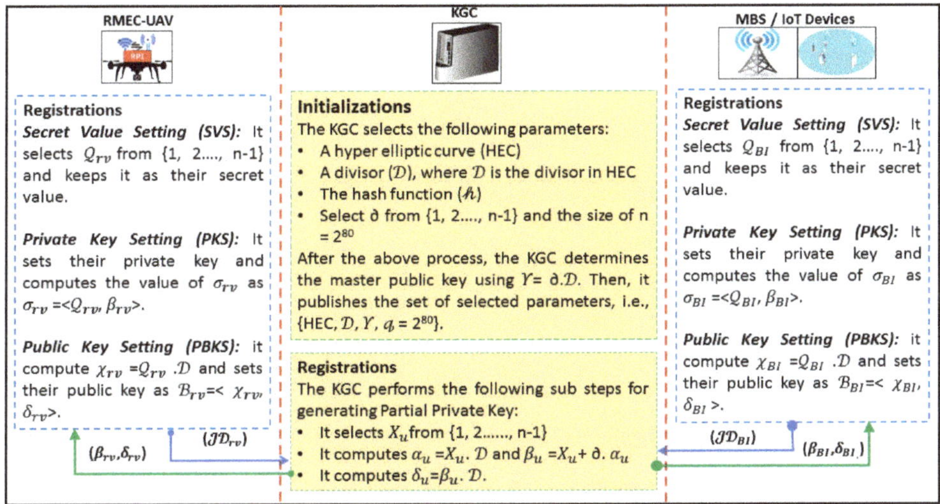

Figure 3. Initialization and registration phase.

5.2. Signing and Verifying Phase

In this part, the *RMEC-UAV* first selects ω from $\{1, 2, \dots, n-1\}$ and then computes $\Delta_1 = \omega/Q_{\mathrm{rv}}$, $\Delta_2 = \beta_{\mathrm{rv}}/\omega$, and then sends it (Δ_1, Δ_2) to the M-UAV. Further, the M-UAV proceeds as follows:

- It selects (τ, φ) from $\{1, 2, \dots, n-1\}$;
- It computes $\mathcal{E} = \hbar(\updownarrow, \Delta_1, \Delta_2, \sigma_{\mathrm{muv}})$ and $\mathcal{Z} = \mathcal{E} + \varphi$;
- It sends $\mathcal{Z}$ to the *RMEC-UAV*. The *RMEC-UAV* generates the partial blind signature $\mathcal{S}^* = Q_{\mathrm{rv}} - \mathcal{Z}.\beta_{\mathrm{rv}}$ and sends it to the M-UAV;
- The M-UAV, then, generates the hash value as $\mathrm{V} = (\updownarrow, \mathrm{Ns})$ and full blind signature, using $\mathcal{S}^{**} = \mathcal{S}^* - \tau$, and transfers it $(\mathcal{S}^{**}, \mathrm{V})$ to the BS/IoT.

Then BS/IoT can verify the blind signature if either of the following equalities are satisfied: $\mathrm{V}^* = (\updownarrow, \mathrm{Ns}) = \mathrm{V} = (\updownarrow, \mathrm{Ns})$ or $\mathrm{V}^* = \mathrm{V}$.

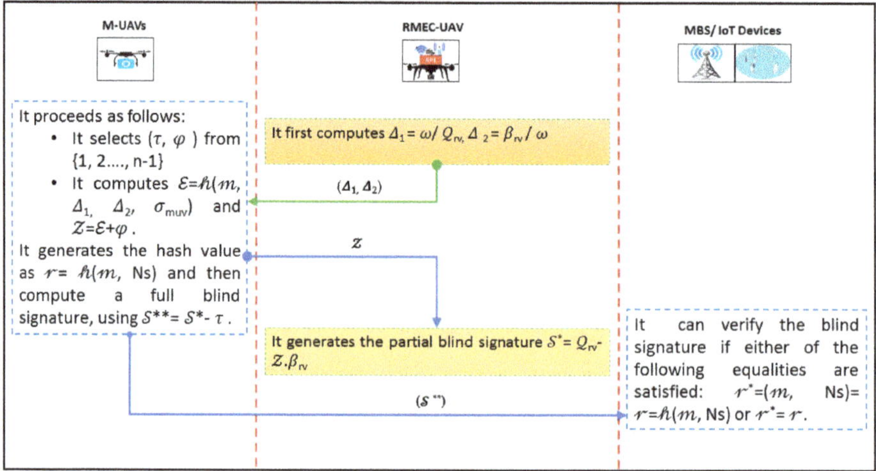

Figure 4. Signing and verifying phase.

6. Security Analysis

This section aims to justify the effectiveness of the proposed scheme in resisting well-known attacks.

6.1. Informal Security Analysis

6.1.1. Theorem 1 ← Unforgeability

A certificateless blind signature is obviously assumed to provide security from a forgeability attack if there is no malicious attacker, $\mathcal{MA}$, which produces the forge blind signature.

Proof. In our case, if an $\mathcal{MA}$ desires the generation of the forge blind signature, then he/she must compute Equation (5). Here, it is the need of S^*, and can be calculated from Equation (6); however, processing this equation, is the need for the calculation of Q_s from Equation (7) is further needed, which is equal to the processing of the hyper elliptic curve discrete logarithm problem. Also, it is a need for β_s from Equation (8), which further requires an equivalent process for the hyper elliptic curve discrete logarithm problem. Thus, the aforementioned assumption proves that an $\mathcal{MA}$ cannot generate the forge blind signature.

$$S^{**} = S^* - \tau, \tag{5}$$

$$S^* = Q_s - Z. \beta_s, \tag{6}$$

$$\sigma_s = Q_s.\mathcal{D}, \tag{7}$$

$$\delta_s = \beta_s.\mathcal{D}. \tag{8}$$

□

6.1.2. Theorem 1 ← Integrity

A certificateless blind signature is supposed to secure from integrity attack if there is no malicious attacker, $\mathcal{MA}$, which becomes modified in the plain text.

Proof. In our proposed work, the requester produces the hash value of a plain text $\updownarrow$ as $X = \hbar_2 (\updownarrow, Ns)$ and sent it to the verifier, along with signature S^{**} and $\updownarrow$ as $(S^{**}, \updownarrow)$. However, if the $\mathcal{MA}$ wishes to change the plain text $\updownarrow$ into $\vec{\updownarrow}$, then the $\mathcal{MA}$ also needs to amend $X = \hbar_2\,(\updownarrow, \text{Ns})$ into $\vec{X} = \hbar_2\,(\vec{\updownarrow}, \text{Ns})$. Therefore, the $\mathcal{MA}$ cannot perform this process because of the one-way nature of the hash function. Thus, keeping in view the aforesaid discussion, our scheme is far more secure against breaking the integrity of plain text. □

6.1.3. Theorem 1 ← Unlinkability

A certificateless blind signature is presumed to offer security from the linkability attack if the blind signer has no access to the plain text.

Proof. In our designed scheme, first of all, the requester selects two blind factors (τ, φ), then performs calculations to find out the value of hash, using $\nabla = \hbar_1(\updownarrow, \Delta_1, \Delta_2, \sigma_r)$, and $\mathcal{Z}$, using $\mathcal{Z} = \nabla + \varphi$. Further, the requester sends $\mathcal{Z}$ to the blind signer. In case the signer wants to see the plain text, it is mandatory for him/her to recover $\updownarrow$ from ∇, where $\nabla = \hbar_1(\updownarrow, \Delta_1, \Delta_2, \sigma_r)$. This, however, is not feasible because of the one-way nature of the hash function. After this, the signer also needs the blind factor φ, which is only known to the requester. Thus, the aforementioned discussion clearly justifies that the scheme, decently, fulfills the security property of unlinkability. □

6.1.4. Theorem 1 ← Replay Attack

In the proposed scheme, the adversary may not give responses to old messages.

Proof. The scheme is resilient against replay attack by offering renewal of nonce Ns. In case an attacker intrudes the message of one session, he/she may not intrude the messages of other sessions with the same Ns because the Ns is renewed at every instance. The receiver is required to perform an up-to-date check with every message and, in the case of an outdatedness being detected in the message, that particular message is trashed into the black box. □

6.2. Formal Security Analysis Using Analysis

In this subsection, results produced from the simulation work using AVISPA tool are presented [56]. This is done, primarily, to ascertain the potency of the proposed scheme against replay and man-in-the-middle attacks. AVISPA is a push-button tool for providing an expressive and modular formal language to simulate protocols and their security properties. SPAN (specific protocol animator for AVISPA) [57], the protocol of security animator for AVISPA, is designed to assist protocol developers write high level protocol specification language (HLPSL) specifications [58]. The HLPSL specifications are interpreted into an intermediate format (IF) by the HLPSLIF translator. Then, it is transformed to the output format (OF) with either on-the-fly model-checker (OFMC) [59], CL-based attack searcher (AtSe) [60], SAT-based model-checker (SATMC), or tree automata-based protocol analyzer (TA4SP). These embedded tools examine the security claims of the aforementioned IF code of an algorithm for two types of attack—replay and man-in-the-middle attacks. The IF code works under two validation states: SAFE, if the cryptographic scheme can safeguard the man-in-the-middle attack, and UNSAFE, in cases where the IF code does not provide protection against man-in-the-middle attack. Formal security verification using the AVISPA tool can be found in several studies to determine the security of many authentication protocols against replay along with man-in-the-middle attacks [61–66]. The basic structure of the AVISPA tool is revealed in Figure 5.

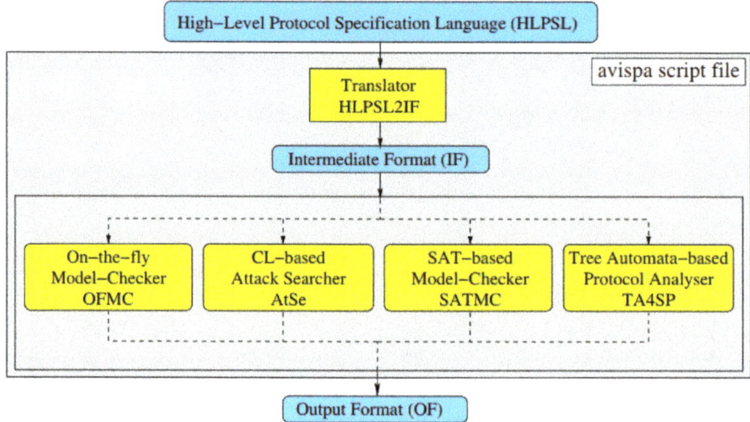

Figure 5. Architecture of the automated validation of internet security protocols and applications (AVISPA) tool v.1.1 [67].

7. Performance Comparison

This section compares the performance of the proposed scheme with the existing counterparts suggested by Lei et al. [4], Islam et al. [47], Nayak et al. [48], and Chen et al. [49].

7.1. Computational Cost

In Table 2, the proposed scheme is compared, in terms of computational cost, with the existing ones, that is, Lei et al.'s scheme [4], Islam et al.'s scheme [47], Nayak et al.'s scheme [48], and Chen et al.'s scheme [49], hereinafter also referred to as the "four chosen schemes", on the basis of major operations. We considered hyperelliptic divisor multiplication as elliptic curve scalar multiplication, and bilinear pairings are the most expensive operations used in the relevant existing schemes. The variables $\hbar\mathfrak{I}$, *em*, $L_{\surd}$ and $\mathfrak{I}\S_{\surd}$ denote the hyperelliptic curve divisor multiplication, elliptic curve scalar multiplication, bilinear pairing, and modular exponential, respectively. It has been observed that a single scalar multiplication takes 0.97 ms for elliptic curve point multiplication (ECPM), 14.90 ms for bilinear pairing, and 1.25 ms for modular exponential [15]. The Multi-Precision Integer and Rational Arithmetic C Library (MIRACL) [68] was used to test the runtime of the basic cryptographic operations up to 1000 times to measure the performance of the proposed approach. The phenomenon was observed on a workstation having following specifications: Intel Core i7- 4510U CPU @ 2.0 GHz, 8 GB RAM and Windows 7 Home Basic 64-bit Operating System [19]. Similarly, the hyperelliptic curve divisor multiplication (HCDM) was assumed to be 0.48 ms due to the smaller key size—80 bit key size [69].

Table 2. Computational cost.

Schemes	Signing	Verifying	Total
Lei et al.'s scheme [4]	16 *em*	5 *em*	21 *em*
Islam et al.'s scheme [47]	7 *em*	$1 L_{\surd} + 4$ *em*	11 *em* $+ 1 L_{\surd}$
Nayak et al.'s scheme [48]	5 *em*	2 *em*	7 *em*
Chen et al.'s scheme [49]	2 *em* $+ 3 \mathfrak{I}\S_{\surd}$	1 *em* $+ 1 L_{\surd} + 1 \mathfrak{I}\S_{\surd}$	3 *em* $+ 1 L_{\surd} + 4 \mathfrak{I}\S_{\surd}$
Proposed	$1 \hbar\mathfrak{I}$	$0 \hbar\mathfrak{I}$	$1 \hbar\mathfrak{I}$

The computational costs provided in Table 3 and illustrated in Figure 6 clearly show that our proposed scheme, when compared with the "four chosen schemes" outperforms in terms of computational cost. The presented scheme is quicker than the existing ones by the following degrees:

Lei et al. [4] by 97.64% ($20.37 - 0.48/20.37 \times 100 = 97.64\%$); Islam et al.'s scheme [47] by 98.12% ($25.57 - 0.48/25.57 \times 100 = 98.12\%$); Nayak et al.'s scheme [48] by 92.93% ($6.79 - 0.48/6.79 \times 100 = 92.93\%$); and Chen et al.'s scheme [49] by 97.89% ($22.81 - 0.48/22.81 \times 100 = 97.89\%$).

Table 3. Computational cost in milliseconds.

Schemes	Signing	Verifying	Total
Lei et al.'s scheme [4]	15.52 ms	4.85 ms	20.37 ms
Islam et al.'s scheme [47]	6.79 ms	18.78 ms	25.57 ms
Nayak et al.'s scheme [48]	4.85 ms	1.94 ms	6.79 ms
Chen et al.'s scheme [49]	5.69 ms	17.12 ms	22.81 ms
Proposed	0.48 ms	0 ms	0.48 ms

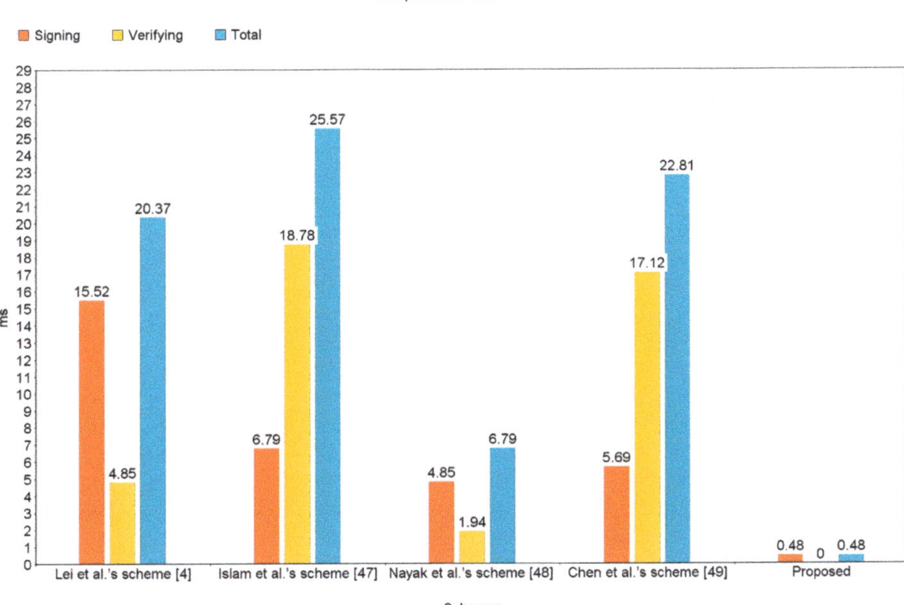

Figure 6. Computational cost (in ms).

7.2. Communication Cost

In this subsection, the proposed scheme is compared, in terms of communication cost, with the existing ones, these being Lei et al.'s scheme [4], Islam et al.'s scheme [47], Nayak et al.'s scheme [48], and Chen et al.'s scheme [49]. For the comparison, we supposed that, $|\mathbb{G}| = 1024$ bits, $|\mathbb{Z}_{LI}| = 160$ bits, $|\mathbb{Z}_n| = 80$ bits, $|H| = 512$ bits, $|\mathbb{C}| = 1024$ bits, and $|\mathcal{W}| = 1024$ bits [70]. According to our suppositions, the communication cost for Lei et al.'s scheme [4] is $4|\mathbb{Z}_{LI}| + 2|\mathcal{W}| + |H| + |\mathbb{C}|$, for Islam et al.'s scheme [47] is $2|\mathbb{G}| + |\mathbb{C}|$, for Nayak et al.'s scheme [48] is $2|\mathbb{Z}_{LI}| + |\mathbb{C}|$, for Chen et al.'s scheme [49] is $|\mathbb{G}| + |H| + |\mathbb{C}|$, and for our proposed scheme is $|\mathbb{Z}_n| + |H| + |\mathbb{C}|$.

The reduction in communication cost of our proposed CL-BS scheme compared with the existing ones as provided in Figure 7 is shown by following degrees: from Lei et al.'s scheme [4] at $(4|\mathbb{Z}_{LI}| + 2|\mathcal{W}| + |H| + |\mathbb{C}|) - (|\mathbb{Z}_n| + |H| + |\mathbb{C}|)/(4|\mathbb{Z}_{LI}| + 2|\mathcal{W}| + |H| + |\mathbb{C}|) = (4224 - 1616/4224 \times 100 = 61.74\%)$; from Islam et al.'s scheme [47] at $(2|\mathbb{G}| + |\mathbb{C}|) - (|\mathbb{Z}_n| + |H| + |\mathbb{C}|)/(2|\mathbb{G}| + |\mathbb{C}|) = (3072 - 1616/3072 \times 100 = 47.39\%)$; from Nayak et al.'s scheme [48] at $(2|\mathbb{Z}_{LI}| + |H| + |\mathbb{C}|) - (|\mathbb{Z}_n| + |H| + |\mathbb{C}|)/(2|\mathbb{Z}_{LI}| + |H| + |\mathbb{C}|) =$

$(1856 - 1616/1856 \times 100 = 14.8\%)$; and from Chen et al.'s scheme [49] at $(|\mathbb{G}| + |H| + |\mathbb{Q}|) - (|\mathcal{Z}_n| + |H| + |\mathbb{Q}|)/(|\mathbb{G}| + |H| + |\mathbb{Q}|) = (2560 - 1616/2560 \times 100 = 36.78\%)$.

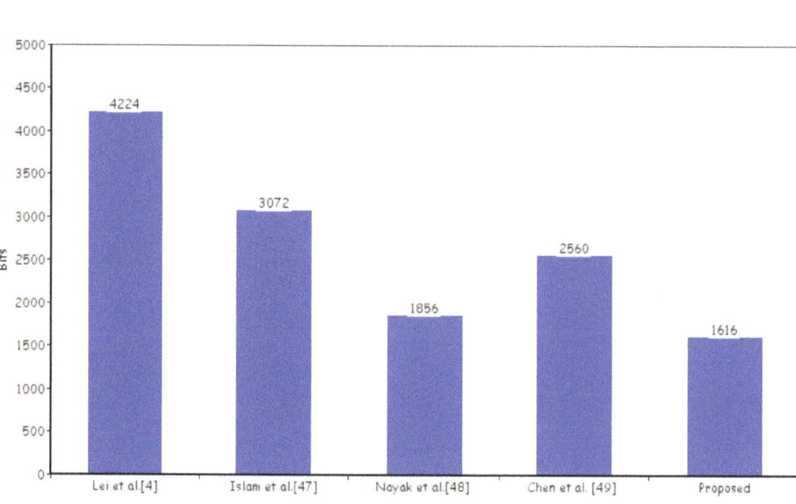

Figure 7. Communication cost (in bits).

7.3. Security Funtionalities

Table 4 presents a brief comparison between the proposed scheme and relevant existing schemes in terms of security functionality. It is worth noting, from Table 4, that the related schemes are not validated through formal security validation tools, such as AVISPA, and none of them guarantee replay attack (RA) and integrity (I). Our proposed scheme is shown to be resistant against various attacks through formal analysis using the widely-accepted automated validation for internet security validation and application (AVISPA) tool as shown in Appendix A.

Table 4. Comparison with relevant existing schemes. Legend: U: unforgeability, I: integrity, UL: unlinkability, RA: replay attack, FA: formal analysis; symbols: ✓: satisfies the security functionality, X: does not satisfy the security functionality.

Schemes	Security Functionalities				
	Informal				Formal
	U	I	UL	RA	FA
Lei et al.'s scheme [4]	✓	X	X	X	X
Islam et al.'s scheme [47]	✓	X	✓	X	X
Nayak et al.'s scheme [48]	✓	X	X	X	X
Chen et al.'s scheme [49]	✓	X	X	X	X
Proposed	✓	✓	✓	✓	✓

8. Conclusions

In this article, we proposed an efficient and provably secure certificateless signature scheme, CL-BS, based on multi-access edge computing (MEC) for a FANET environment using the concept of hyperelliptic curve. The proposed scheme was shown to be resistant against various attacks through informal security analysis, as well as through the formal security verification using the widely-accepted

AVISPA tool. The scheme was also efficient in terms of computational and communication costs. On doing a comparative analysis with existing counterparts, it was noticed that the proposed scheme was characterized by least computational and communication costs, these being 0.48 ms and 1616 bits, respectively, which authenticates the superiority of our scheme.

In future, we intend to integrate a computational offloading and scheduling mechanism, where M-UAVs will offload and schedule the computing tasks in the RMEC-UAV for fast processing and execution.

Author Contributions: Conceptualization, M.A.K. and I.M.Q.; methodology and implementation, M.A.K., I.M.Q., I.U., and F.N.; simulation, M.A.K. and I.U.; validation, M.A.K., I.M.Q., I.U., and S.K.; data curation, M.A.K., S.K., and F.K.; writing—original draft preparation, M.A.K., F.K., and F.N.; writing—review and editing, M.A.K., F.N., and F.K.; supervision, I.M.Q. All authors have read and agreed to the published version of the manuscript.

Funding: This research received no external funding.

Conflicts of Interest: The authors declare no conflict of interest.

Appendix A. Implementation of Our Proposed CL-BS Scheme in AVISPA

The proposed scheme has been implemented for blind signer and verifier in HLPSL, as illustrated in Algorithms A1 and A2. The experiment was performed on a computer workstation having the specifications as follows: Haier Win8.1 PC; Intel Core i3-4010U CPU @ 1.70 GHz; 64-bit operating system and ×64-based processor. The software platforms consulted were Oracle VM Virtual Box (version: 5.2.0.118431) and SPAN (version: SPAN-Ubuntu-10.10-light_1). As with any security protocol, to be analyzed in AVISPA, the roles for session, goal, and environment were executed as shown in Algorithms A3 and A4. In order to gauge the probability of attacks on the proposed scheme, the widely-used OFMC and CL-AtSe backends were selected for the execution test. Because other backends such as SATMC and TA4SP are not compatible with bitwise XOR operations, the simulation results of SATMC and TA4SP were not included in the research work. Here, it is imperative to ascertain the execution of specified protocol in terms of whether the authentic agents can execute the specified protocol or not. To do so, the back-ends perform check operations. Then, the information is provided to the intruder about a few normal sessions between authentic agents. Secondly, the susceptibility of the system to man-in-the-middle attack is also estimated by the back-ends. This is done to verify the Dolev–Yao (DY) model [54]. The scheme is also simulated under SPAN (specific protocol animator for AVISPA) web-tool. The results for OFMC and AtSe are shown in Figures A1 and A2, respectively. It is evident that the scheme is safe against replay and man-in-the-middle attack.

Algorithm A1 High-level protocol specification language (HLPSL) code for Signer role

---

```
role
role_Blindsigner(Blindsigner:agent,Verifier:agent,Xs:public_key,Xv:public_key,SND,RCV:channel(dy))
played_by Blindsigner
def=
local
State:nat,Ns:text,Sub:hash_func,Z:text,T:text
init
State: = 0
transition
1. State=0 /\ RCV (start) =|> State': =1 /\
SND (Blindsigner.Verifier)
2. State=1 /\ RCV (Verifier. {Ns'} _Xv) =|> State': =2 /\ T':=new() /\ Z':=new() /\
SND(Blindsigner.{Sub(Z'.T')}_inv(Xs))
end role
```

Algorithm A2 High-level protocol specification language (HLPSL) code for Verifier role

---

role
role_Verifier(Blindsigner:agent,Verifier:agent,Xs:public_key,Xv:public_key,SND,RCV:channel(dy))
played_by Verifier
def=
local
State:nat,Ns:text,Sub:hash_func,Z:text,T:text
init
State: = 0
transition
1. State=0 /\ RCV(Blindsigner.Verifier) =|> State':=1 /\ Ns':=new() /\ SND(Verifier.{Ns'}_Xv)
2. State=1 /\ RCV (Blindsigner. {Sub (Z'. T')} _inv (Xs)) =|> State': =2
end role

Algorithm A3 High-level protocol specification language (HLPSL) code for Sessions role

---

role
session1(Blindsigner:agent,Verifier:agent,Xs:public_key,Xv:public_key)
def=
local
SND2,RCV2,SND1,RCV1:channel(dy)
composition
role_Verifier(Blindsigner,Verifier,Xs,Xv,SND2,RCV2) /\
role_Blindsigner(Blindsigner,Verifier,Xs,Xv,SND1,RCV1)
end role

role
session2(Blindsigner:agent,Verifier:agent,Xs:public_key,Xv:public_key)
def=
local
SND1,RCV1:channel(dy)
composition
role_Blindsigner(Blindsigner,Verifier,Xs,Xv,SND1,RCV1)
end role

Algorithm A4 High-level protocol specification language (HLPSL) code for Environment role

---

role
environment()
def=
const

hash_0:hash_func,xs:public_key,alice:agent,bob:agent,xv:public_key,const_1:agent,const_2:public_key,
const_3:public_key,auth_1:protocol_id,sec_2:protocol_id
intruder_knowledge = {alice,bob}
composition
session2(i, const_1, const_2, const_3) /\ session1(alice,bob,xs,xv)
end role

goal
authentication_on auth_1
secrecy_of sec_2
end goal

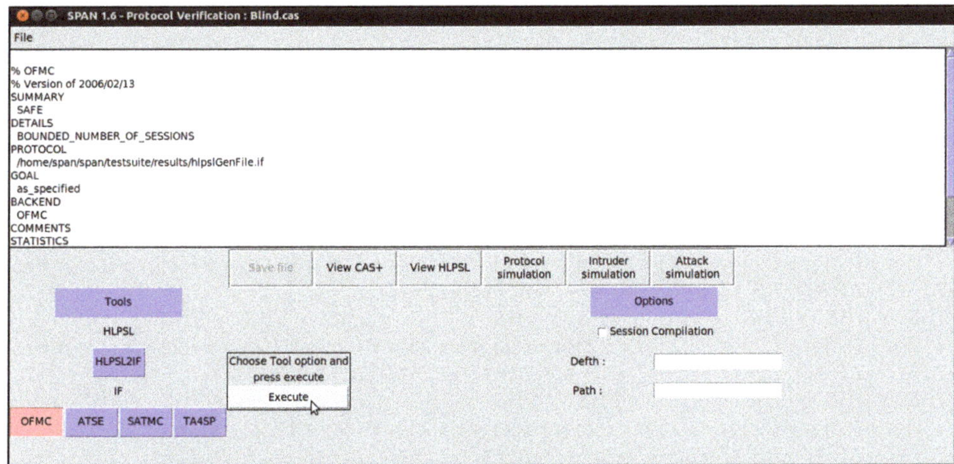

Figure A1. Simulation results for on-the-fly model-checker (OFMC).

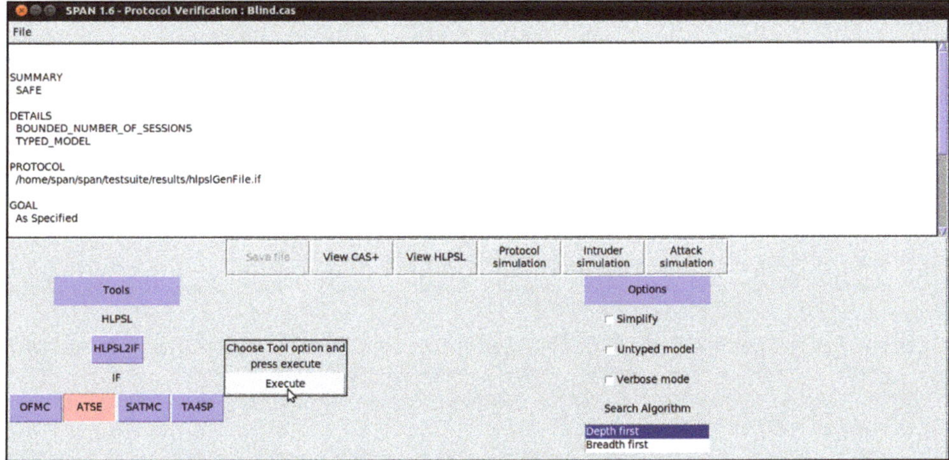

Figure A2. Simulation results for AtSe.

References

1. Khan, M.A.; Qureshi, I.M.; Khanzada, F.A. Hybrid Communication Scheme for Efficient and Low-Cost Deployment of Future Flying Ad-Hoc Network (FANET). *Drones* **2019**, *3*, 16. [CrossRef]
2. Bekmezci, I.; Sahingoz, O.K.; Temel, Ş. Flying ad-hoc Network (FANET): A survey. *Ad Hoc Netw.* **2013**, *11*, 1254–1270. [CrossRef]
3. Khan, M.A.; Safi, A.; Qureshi, I.M.; Khan, I.U. Flying ad-hoc Network (FANET): A review of communication architectures, and routing protocols. In Proceedings of the 2017 First International Conference on Latest trends in Electrical Engineering and Computing Technologies (INTELLECT), Karachi, Pakistan, 15–16 November 2017; pp. 1–9.
4. He, L.; Ma, J.; Mo, R.; Wei, D. Designated Verifier Proxy Blind Signature Scheme for Unmanned Aerial Vehicle Network Based on Multi-access Edge Computing (MEC). *Secur. Commun. Netw.* **2019**, *2019*, 8583130. [CrossRef]

5. Khan, M.A.; Qureshi, I.M.; Khan, I.U.; Nasim, M.A.; Javed, U.; Khan, M.W. On the performance of flying ad-hoc Network (FANET) with directional antennas. In Proceedings of the 2018 5th International Multi-Topic ICT conference (IMTIC), Jamshoro, Pakistan, 25–27 April 2018; pp. 1–8.

6. Khan, M.A.; Khan, I.U.; Safi, A.; Quershi, I.M. Dynamic Routing in Flying Ad-Hoc Networks Using Topology-Based Routing Protocols. *Drones* **2018**, *2*, 27. [CrossRef]

7. Suárez-Albela, M.; Fraga-Lamas, P.; Fernández-Caramés, T.M. A Practical Evaluation on RSA and ECC-Based Cipher Suites for IoT High-Security Energy-Efficient Fog and Mist Computing Devices. *Sensors* **2018**, *18*, 3868. [CrossRef]

8. Yu, M.; Zhang, J.; Wang, J.; Gao, J.; Xu, T.; Deng, R.; Zhang, Y.; Yu, R. Internet of Things security and privacy-preserving method through nodes differentiation, concrete cluster centers, multi-signature, and blockchain. *Int. J. Distrib. Sens. Netw.* **2018**, *14*, 12. [CrossRef]

9. Braeken, A. PUF Based Authentication Protocol for IoT. *Symmetry* **2018**, *10*, 8. [CrossRef]

10. Zhou, C.; Zhao, Z.; Zhou, W.; Mei, Y. Certificateless Key-Insulated Generalized Signcryption Scheme without Bilinear Pairings. *Secur. Commun. Netw.* **2017**, *2017*, 8405879. [CrossRef]

11. Kumari, S.; Karuppiah, M.; Das, A.K.; Li, X.; Wu, F.; Kumar, N. A secure authentication scheme based on elliptic curve cryptography for IoT and cloud servers. *J. Supercomput.* **2017**, *74*, 12. [CrossRef]

12. Omala, A.; Mbandu, A.; Mutiria, K.; Jin, C.; Li, F. Provably Secure Heterogeneous Access Control Scheme for Wireless Body Area Network. *J. Med. Syst.* **2018**, *42*, 6. [CrossRef]

13. Tamizhselvan, C.; Vijayalakshmi, V. An Energy Efficient Secure Distributed Naming Service for IoT. *Int. J. Adv. Stud. Sci. Res.* **2019**, *3*, 8.

14. Naresh, V.S.; Sivaranjani, R.; Murthy, N.V.E.S. Provable secure lightweight hyper elliptic curve-based communication system for wireless sensor Network. *Int. J. Commun. Syst.* **2018**, *31*, 15. [CrossRef]

15. Rahman, A.; Ullah, I.; Naeem, M.; Anwar, R.; Khattak, H.S.; Ullah, A. Lightweight Multi-Message and Multi-Receiver Heterogeneous Hybrid Signcryption Scheme based on Hyper Elliptic Curve. *Int. J. Adv. Comput. Sci. Appl.* **2018**, *9*, 5. [CrossRef]

16. Won, J.; Seo, S.H.; Bertino, E. Certificateless Cryptographic Protocols for Efficient Drone-Based Smart City Applications. *IEEE Access* **2017**, *5*, 3721–3749. [CrossRef]

17. Barka, E.; Kerrache, C.; Hussain, R.; Lagraa, N.; Lakas, A.; Bouk, S. A Trusted Lightweight Communication Strategy for Flying Named Data Networking. *Sensors* **2018**, *18*, 2683. [CrossRef]

18. Bae, M.; Kim, H. Authentication and Delegation for Operating a Multi-Drone System. *Sensors* **2019**, *19*, 2066. [CrossRef]

19. Seo, S.-H.; Won, J.; Bertino, E. pCLSC-TKEM: A pairing-free certicateless signcryption-tag key encapsulation mechanism for a privacy-preserving IoT. *Trans. Data Priv.* **2016**, *9*, 101–130.

20. Liu, W.; Strangio, M.A.; Wang, S. Efficient Certificateless Signcryption Tag-KEMs for Resource constrained Devices. *arXiv* **2015**, *1510*, 01446.

21. Reddy, P.V.; Babu, A.R.; Gayathri, N.B. Efficient and Secure Identity-based Strong Key Insulated Signature Scheme without Pairings. *J. King Saud Univ. Comput. Inf. Sci.* **2018**.

22. Xiong, H.; Mei, Q.; Zhao, Y. Efficient and provably secure certificateless parallel key-insulated signature without pairing for IIoT environments. *IEEE Syst. J.* **2019**, 1–11. [CrossRef]

23. Bekkouche, O.; Taleb, T.; Bagaa, M. UAVs Traffic Control based on Multi-Access Edge Computing. In Proceedings of the 2018 IEEE Global Communications Conference (GLOBECOM 2018), Abu Dhabi, UAE, 9–13 December 2018.

24. Ouahouah, S.; Taleb, T.; Song, J.; Benzaid, C. Efficient offloading mechanism for UAVs-based value-added services. In Proceedings of the 2017 IEEE International Conference on Communications (ICC), Paris, France, 21–25 May 2017; pp. 1–6.

25. Motlagh, N.H.; Bagaa, M.; Taleb, T. Uav-based iot platform: A crowd surveillance use case. *IEEE Commun. Mag.* **2017**, *55*, 128–134. [CrossRef]

26. ETSI. Multi-Access Edge Computing (MEC). In *Framework and Reference Architecture*; DGS MEC; ETSI: Sophia Antipolis, France, 2016; Volume 3.

27. Garg, S.; Singh, A.; Batra, S.; Kumar, N.; Yang, L.T. UAV empowered edge computing environment for cyber-threat detection in smart vehicles. *IEEE Netw.* **2018**, *32*, 42–51. [CrossRef]

28. Lee, J.; Lee, J. Hierarchical Multi-access Edge Computing (MEC) architecture based on context awareness. *Appl. Sci.* **2018**, *8*, 1160. [CrossRef]

29. Intharawijitr, K.; Iida, K.; Koga, H. Simulation study of low latency network architecture using Multi-access Edge Computing (MEC). *IEICE Trans. Inf. Syst.* **2017**, *E100D*, 963–972. [CrossRef]
30. Messous, M.-A.; Sedjelmaci, H.; Houari, N.; Senouci, S.-M. Computation offloading game for an UAV network in Multi-access Edge Computing (MEC). In Proceedings of the 2017 IEEE International Conference on Communications, ICC 2017, Paris, France, 21–25 May 2017; pp. 1–6.
31. Ansari, N.; Sun, X. Multi-access Edge Computing (MEC) empowers internet of things. *IEICE Trans. Commun.* **2018**, *E101B*, 604–619. [CrossRef]
32. Zhang, K.; Leng, S.; He, Y.; Maharjan, S.; Zhang, Y. Multi-access Edge Computing (MEC) and networking for green and low-latency internet of things. *IEEE Commun. Mag.* **2018**, *56*, 39–45. [CrossRef]
33. Grasso, C.; Schembra, G. A Fleet of MEC UAVs to Extend a 5G Network Slice for Video Monitoring with Low-Latency Constraints. *J. Sens. Actuator Netw.* **2019**, *8*, 3. [CrossRef]
34. Chaum, D. Blind signatures for untraceable payments. *Adv. Cryptol.* **1983**, 199–203.
35. Mambo, M.; Usuda, K.; Okamoto, E. Proxy signatures for delegating signing operation. In Proceedings of the 3rd ACM Conference on Computer and Communications Security, New Delhi, India, 14–15 March 1996; pp. 48–56.
36. Tan, Z.; Liu, Z.; Tang, C. Digital proxy blind signature schemes based on DLP and ECDLP. *MM Res. Prepr.* **2002**, *21*, 212–217.
37. Tan, Z. Efficient pairing-free provably secure identity-based proxy blind signature scheme. *Secur. Commun. Netw.* **2013**, *6*, 593–601. [CrossRef]
38. Yang, F.-Y.; Liang, L.-R. A proxy partially blind signature scheme with proxy revocation. *J. Ambient Intell. Humaniz. Comput.* **2013**, *4*, 255–263. [CrossRef]
39. Verma, G.K.; Singh, B.B. Efficient message recovery proxy blind signature scheme from pairings. *Trans. Emerg. Telecommun. Technol.* **2017**, *28*, e3167. [CrossRef]
40. Zhu, H.; Tan, Y.-A.; Zhu, L.; Zhang, Q.; Li, Y. An efficient identity-based proxy blind signature for semi offline services. *Wirel. Commun. Mob. Comput.* **2018**, *2018*, 5401890. [CrossRef]
41. Jakobsson, M.; Sako, K.; Impagliazzo, R. Designated verifier proofs and their applications. In Proceedings of the International Conference on the Theory and Applications of Cryptographic Techniques, Berlin, Heidelberg, 13 July 2001; pp. 143–154.
42. Dai, J.Z.; Yang, X.H.; Dong, J.X. Designated-receiver proxy signature scheme for electronic commerce. In Proceedings of the IEEE International Conference on Systems, Man and Cybernetics, Washington, DC, USA, 10 November 2003; pp. 384–389.
43. Huang, X.; Mu, Y.; Susilo, W.; Zhang, F. Short designated verifier proxy signature from pairings. In Proceedings of the International Conference on Embedded and Ubiquitous Computing, Nagasaki, Japan, 6–9 December 2005; pp. 835–844.
44. Shim, K.-A. Short designated verifier proxy signatures. *Comput. Electr. Eng.* **2011**, *37*, 180–186. [CrossRef]
45. Islam, S.H.; Biswas, G. A provably secure identity-based strong designated verifier proxy signature scheme from bilinear pairings. *J. King Saud Univ. Comput. Inf. Sci.* **2014**, *26*, 55–67. [CrossRef]
46. Hu, X.; Tan, W.; Xu, H.; Wang, J. Short and provably secure designated verifier proxy signature scheme. *IET Inf. Secur.* **2016**, *10*, 69–79. [CrossRef]
47. Islam, S.; Obaidat, M.S. Design of provably secure and efficient certificateless blind signature scheme using bilinear pairing. *Secur. Commun. Netw.* **2015**, *8*, 4319–4332. [CrossRef]
48. Nayak, S.K.; Mohanty, S.; Majhi, B. CLB-ECC: Certificateless Blind Signature Using ECC. *JIPS* **2017**, *13*, 970–986.
49. Chen, H.; Zhang, L.; Xie, J.; Wang, C. New Efficient Certificateless Blind Signature Scheme. In Proceedings of the 2016 IEEE Trustcom/BigDataSE/ISPA, Tianjin, China, 23–26 August 2016; pp. 349–353.
50. Koblitz, N. Elliptic curve cryptosystems. *Math. Comput.* **1987**, *48*, 203–209. [CrossRef]
51. Hyperelliptic Curve. 2019. Available online: https://en.wikipedia.org/wiki/Hyperelliptic_curve (accessed on 25 October 2019).
52. Pelzl, J.; Wollinger, T.; Guajardo, J.; Paar, C. Hyperelliptic curve cryptosystems: Closing the performance gap to elliptic curves. In *International Workshop on Cryptographic Hardware and Embedded Systems*; Springer: Berlin, Heidelberg, 2003; pp. 351–365.
53. Cantor, D.G. Computing in Jacobian of a Hyperelliptic Curve. *Math. Comput.* **1987**, *48*, 95–101. [CrossRef]
54. Dolev, D.; Yao, A. On the security of public key protocols. *IEEE Trans. Inf. Theory* **1983**, *29*, 198–208. [CrossRef]

55. Siddiqi, M.A.; Yu, H.; Joung, J. 5G Ultra-Reliable Low-Latency Communication Implementation Challenges and Operational Issues with IoT Devices. *Electronics* **2019**, *8*, 981. [CrossRef]
56. AVISPA. Automated Validation of Internet Security Protocols and Applications. 2019. Available online: http://www.avispa-project.org/ (accessed on 25 October 2019).
57. AVISPA. SPAN: A Security Protocol Animator for AVISPA. 2019. Available online: http://www.avispa-project.org/ (accessed on 25 October 2019).
58. Oheimb, D.V. The high-level protocol specification language HLPSL developed in the EU project avispa. In Proceedings of the APPSEM 2005 Workshop, Tallinn, Finland, 13–15 September 2005; pp. 1–2.
59. Basin, D.; Modersheim, S.; Vigano, L. OFMC: A symbolic model checker for security protocols. *Int. J. Inf. Secur.* **2005**, *4*, 181–208. [CrossRef]
60. Turuani, M. The CL-Atse porotocol analyser. In Proceedings of the International Coneference on Rewriting Techniques and Applications (RTA), Seattle, WA, USA, 12–14 August 2006; pp. 227–286.
61. Yu, S.; Lee, J.; Lee, K.; Park, K.; Park, Y. Secure Authentication Protocol for Wireless Sensor Network in Vehicular Communications. *Sensors* **2018**, *18*, 3191. [CrossRef]
62. Park, K.; Park, Y.; Park, Y.; Reddy, A.G.; Das, A.K. Provably secure and efficient authentication protocol for roaming service in global mobility Network. *IEEE Access* **2017**, *5*, 25110–25125. [CrossRef]
63. Odelu, V.; Das, A.K.; Choo, K.R.; Kumar, N.; Park, Y.H. Efficient and secure time-key based single sign-on authentication for mobile devices. *IEEE Access* **2017**, *5*, 27707–27721. [CrossRef]
64. Odelu, V.; Das, A.K.; Kumari, S.; Huang, X.; Wazid, M. Provably secure authenticated key agreement scheme for distributed mobile cloud computing services. *Futuer Generat. Comput. Syst.* **2017**, *68*, 74–88. [CrossRef]
65. Park, K.; Park, Y.; Park, Y.; Das, A.K. 2PAKEP: Provably Secure and Efficient Two-Party Authenticated Key Exchange Protocol for Mobile Environment. *IEEE Access* **2018**, *6*, 30225–30241. [CrossRef]
66. Banerjee, S.; Odelu, V.; Das, A.K.; Chattopadhyay, S.; Kumar, N.; Park, Y.H.; Tanwar, S. Design of an Anonymity-Preserving Group Formation Based Authentication Protocol in Global Mobility Network. *IEEE Access* **2018**, *6*, 20673–20693. [CrossRef]
67. AVISPA v1.1 User Manual. 2019. Available online: http://www.avispa-project.org/package/user-manual.pdf (accessed on 25 October 2019).
68. Shamus Sofware Ltd. Miracl library. Available online: http://github.com/miracl/MIRACL (accessed on 25 October 2019).
69. Ullah, I.; Amin, N.U.; Naeem, M.; Khattak, S.; Khattak, S.J.; Ali, H. A Novel Provable Secured Signcryption Scheme PSSS: A Hyper-Elliptic Curve-Based Approach. *Mathematics* **2019**, *7*, 686. [CrossRef]
70. Ullah, I.; Alomari, A.; Ul Amin, N.; Khan, M.A.; Khattak, H. An Energy Efficient and Formally Secured Certificate-Based Signcryption for Wireless Body Area Network with the Internet of Things. *Electronics* **2019**, *8*, 1171. [CrossRef]

Article

A Traceable and Privacy-Preserving Authentication for UAV Communication Control System

Chin-Ling Chen [1,2,3], Yong-Yuan Deng [3,*], Wei Weng [1,*], Chi-Hua Chen [4,*], Yi-Jui Chiu [5] and Chih-Ming Wu [6]

[1] College of Computer and Information Engineering, Xiamen University of Technology, Xiamen 361024, China; clc@mail.cyut.edu.tw

[2] School of Information Engineering, Changchun Sci-Tech University, Changchun 130600, China

[3] Department of Computer Science and Information Engineering, Chaoyang University of Technology, Taichung 41349, Taiwan

[4] College of Mathematics and Computer Science, Fuzhou University, Fuzhou 350116, China

[5] School of Mechanical and Automotive Engineering, Xiamen University of Technology, Xiamen 361024, China; chiuyijui@xmut.edu.cn

[6] School of Civil Engineering and Architecture, Xiamen University of Technology, Xiamen 361024, China; chihmingwu@xmut.edu.cn

* Correspondence: allen.nubi@gmail.com (Y.-Y.D.); wwweng@xmut.edu.cn (W.W.); chihua0826@gmail.com (C.-H.C.)

Received: 15 November 2019; Accepted: 20 December 2019; Published: 1 January 2020

Abstract: In recent years, the concept of the Internet of Things has been introduced. Information, communication, and network technology can be integrated, so that the unmanned aerial vehicle (UAV) from consumer leisure and entertainment toys can be utilized in high value commercial, agricultural, and defense field applications, and become a killer product. In this paper, a traceable and privacy-preserving authentication is proposed to integrate the elliptic curve cryptography (ECC), digital signature, hash function, and other cryptography mechanisms for UAV application. For sensitive areas, players must obtain flight approval from the ground control station before they can control the UAV in these areas. The traditional cryptography services such as integrity, confidentiality, anonymity, availability, privacy, non-repudiation, defense against DoS (Denial-of-Service) attack, and spoofing attack can be ensured. The feasibility of mutual authentication was proved by BAN logic. In addition, the computation cost and the communication cost of the proposed scheme were analyzed. The proposed scheme provides a novel application field.

Keywords: UAV; Mutual authentication; Privacy; Traceable; BAN logic

---

1. Introduction

With the development of battery power, sensing systems, artificial intelligence and other technologies, small commercial unmanned aerial vehicles (UAVs) combining these technologies have, in recent years, become a very popular product. Small UAVs have tremendous potential in different fields and tasks, and have great flexibility in application. In addition to personal aerial photography, entertainment, and commercial markets, they can be used in various monitoring work such as disaster relief [1], in various environments involving animals and plants, coasts and borders [2,3], in freight transportation, military and police law enforcement tasks, and even agricultural and industrial applications [4–8]. Nader et al. [9] pointed out that UAVs could be employed in different ways to achieve smart city services. For example, using UAVs for traffic monitoring and management, merchandise delivery, health and emergency services, and air taxi services can enhance these services in terms of quality, productivity, timeliness, reliability, and performance and could help reduce the costs of offering these services. However, small UAVs also can pose a variety of security threats under improper use.

Although every case of an unmanned aerial vehicle being improperly used has complex security implications, it is difficult to sum this up as a single security threat; for example, in the protection of important persons, unmanned aerial vehicles may violate their privacy, launch attacks, threaten their lives, or destroy their facilities. Different threats in several different cases are examined below.

(1) Personal safety of specific persons and military and police officers: The small UAVs used by the fighters of the Organization of Islamic States could, for example, be used to attack enemy soldiers on the battlefield in the Middle East. This situation can be described as the future personal safety protection work for important persons and law enforcement officers. It is necessary to take precautions against small UAVs.

(2) Protection of key infrastructure: In July 2018, Greenpeace posted a video on the Internet showing the small UAVs operated by members of Greenpeace, painted superhuman, hitting a spent fuel facility near Lyon, France. This incident still reminds us of the importance of UAV protection for key infrastructure or the environment (for example, forest fire detection).

(3) Flight safety: In late December 2018 and early January 2019, London's Gatwick and Heathrow airports were disrupted by UAVs, causing chaos in takeoff, landing, and scheduling. The former even closed for 33 h and cancelled hundreds of flights, causing losses of more than 50 million pounds.

(4) Privacy and confidentiality protection: UAVs can be used to steal important confidential information, such as in Northern Ireland in August 2016, when UAVs were used to take pictures of people entering passwords in ATMs. Small UAVs can even be used as hackers' tools to further steal business secrets. According to the reports of The Times on 21 January 2019, in recent years, secrets have been stolen by eavesdropping, or even masquerading as wireless network connections to obtain employee passwords, etc. More and more companies are seeking anti-UAV technology to ensure against commercial benefits by stealing secrets by disguising wireless network connections to obtain employee passwords and other information.

(5) Other criminal behaviors: In addition to the use of military and police personnel to monitor and assist in law enforcement, small UAVs may also operate in the hands of criminals. The surveillance functions provided by UAVs also enable criminal groups to detect and monitor their targets before committing crimes.

(6) Security loopholes become a hidden concern: In addition to the improper use by the users themselves, UAVs may also be attacked by intentional hackers. By means of security loopholes including GPS and control signals, wireless networks and so on, "hijacking" may take control of a UAV. Vulnerabilities in the UAV manufacturer's security may also become another type of drone-derived security problem. A well-known software technology website Check Point reported in November 2018 that the world's largest manufacturer, China's Dajiang, has a security vulnerability in its identity authentication process. If it is attacked by a hacker, it may leak the location of the operator and the captured image, etc. Even the possibility of intercepting the carried goods also highlights the security problems of drones.

To sum up, in spite of UAVs being widely used in civilian, commercial, and military applications in recent years, because they use wireless networks for information exchange, there are many security issues that are faced.

Firstly, "privacy" refers to the part of an individual that he does not want to be known by others, and that he has the right to protect. In English, "to be let alone" means to "not be disturbed by others", which is the basic spirit of privacy. Privacy also means "secret". In general, what we call privacy refers to information privacy. Privacy and freedom are related to individual behavior rather than inappropriate observation and interference by others. The interests of privacy include sexual activities, religious practices, and political activities. What is the importance of privacy? Privacy is about human dignity, personal subjectivity, and personality development. If some of a person's own information is exposed, he will feel uncomfortable, embarrassed, or harassed by others, and it will be difficult to live comfortably. Compared with personal privacy, sensitive information of the state or government has a greater impact.

Secondly, the malicious attacker can perform passive eavesdropping, active interfering, leaking of secret information, data tampering, denial of service, message misuse, message replay, and impersonation attack between sender and receiver. This will cause the resource collapse attack, and even disturb the operations of routing protocol for UAVs [10]. UAVs are conducted in flying ad hoc networks (FANETs) which should provide defense against various known attacks under wireless environment.

Thirdly, because of the specific properties of FANET (wireless links, collaborative characteristics, uncontrollable environment, and lack of a fixed infrastructure) securing the network is difficult. The traditional security issues are availability, authentication, integrity, and confidentiality, which have become targets that the attacker wants to break. [11]. Legitimate UAVs suffer from malicious UAVs by implanting the incorrect information into their sensors. Therefore, it causes these compromised UAVs to transmit the wrong messages for the base station, and thereby endangering the data integrity [10].

In order to legalize and guarantee the privacy of the broadcasted messages, much literature is focused on this issues. For example, Strohmeier et al. [12] surveyed an automatic dependent surveillance-broadcast protocol (ADS-B), and that is an on-board component part of the UAV system, and discussed and listed the vulnerabilities in ADS-B protocol. Wesson et al. [13] further analyzed and evaluated the cryptographic strategies of ADS-B based on their effectiveness and practicality in the cost-averse, technologically-complex, and interoperability-focused aviation community. The purpose of these works was to find a suitable mechanism to ensure the security of the UAVs system for sensitive control areas.

In past literature, some articles [10,14–16] refer to malicious attacks on UAV applications, such as intrusion detection, enhancing security against the lethal cyber-attacks for UAV networks. Therefore, a Q-learning-based UAV power allocation strategy combining Q-learning and deep learning to accelerate the learning speed for attack modes was proposed by Xiao et al. [17]. García-Magariño et al. [16] used a secure asymmetric encryption with a pre-shared list of official UAVs and an agent-based approach to detect if an official UAV is physically hijacked. However, these articles only focus on the intrusion detection or the problem of UAVs being physical hijacked. It is a fact that to prevent all intrusions from being attacked by hackers, the fundamental solution is to propose an effective and comprehensive security protocol. Such a secure mechanism should comprehensively detect and provide information and identity authentication to achieve the purposes of availability, privacy, and non-repudiation and to defend against known attacks for the UAV's environment.

Recently, some literature [18–21] has used specific cryptographic algorithms to implement security mechanisms in UAVs. In 2017, Yoon et al. [18] used the Raspberry Pi to present a design of a second channel security system that can regain control of a UAV when there is an attack on the UAV. In this scheme, the authors only used flow charts to describe the scenario. The authors claimed that they can provide authentication with the ground station and defense against the DoS attack. However, this scheme does not present the detail cryptography scenario and no performance analysis.

Later, Chen et al. [19] proposed a mutual authentication improvement in security. In order to achieve higher efficiency and reduce the computational cost, thus the proposed scheme conformed to the network-connected UAV communication systems, and that satisfied the requirements of the limited bandwidth and computation resources. However, the authors used the asymmetric bilinear pairings mechanism and the cost of this was high and it was not supported by formal proof. Wazid et al. [20] also presented a lightweight remote user authentication and key agreement scheme to solve security issues between the user and the accessed drone in Internet of Drones (IoD) applications.

Recently, Tian et al. [21] proposed an efficient privacy-preserving authentication framework for the edge-assisted Internet of Drones. They followed a predictive UAV authentication approach. The authors considered that location, identity, and flying routes of each legitimate UAV are sensitive information in the IoD network. Therefore, they proposed a secure authentication and privacy protection for an efficient MEC-assisted (mobile edge computing) framework. But this scheme did not consider mutual authentication for ensuring the communication entity.

In fact, due to the UAV's characteristics, it is hard to prevent a privacy leak. Therefore, this study aims to focus on sensitive areas (for example: airports and military areas) to set up this management system and use ECC (elliptic curve cryptography) technology [22,23] to ensure data integrity and nonrepudiation. It is a fact that any intruders can break through the defense function of the system if the security mechanism of the system is not perfect and the user's identity is not authenticated accurately. This study also intends to employ the proof mode of BAN logic mechanism for mutual authentication to eliminate the intrusive chances of malicious attackers.

The paper is organized as follows. The applied mechanisms and security mechanisms are reviewed and discussed in Section 2. The designs and flows of the proposed scheme are presented in Section 3. Security analyses and comparisons are discussed in Section 4. Finally, in Section 5, conclusions are offered.

2. Preliminary and Security Requirements

This section includes two subsections: (1) the elliptic curve cryptography and Diffie–Hellman key exchange are presented in Section 2.1 and (2) security requirements are defined in Section 2.2.

2.1. Elliptic Curve Cryptography and Diffie–Hellman Key Exchange

Elliptic curve cryptography [22,23] was proposed in 1995. Digital signature schemes can be used to provide the following basic cryptographic services: data integrity, data origin authentication, and non-repudiation.

The Diffie–Hellman key exchange [24] is a method for securely exchanging cryptographic keys over a public channel. It is one of the earliest practical examples of public key exchange implemented within the field of cryptography. The Diffie–Hellman key exchange method allows two parties that have no prior knowledge of each other to jointly establish a shared secret key over an insecure channel. This key can then be used to encrypt subsequent communications by using a symmetric key cipher.

The following problems exist for the Elliptic Curve Diffie-Hellman method:

Computational Diffie–Hellman (CDH) Problem: Given aP and bP, where $a, b \in R$, $Z * q$, and P are the generator of G, compute abP.

Decisional Diffie–Hellman (DDH) Problem: Given aP, bP, and cP, where $a, b, c \in R$, $Z * q$, and P are the generators of G, confirm whether or not $cP = abP$, which is equal to confirming whether or not $c = ab \bmod q$.

2.2. Security Requirements

A UAV communication control system has the following main security requirements and known attacks [11,13–15,19,20,25]:

- Mutual authentication: this ensures that only legitimate parties are allowed to participate in the UAV network. There are two types of authentication services: node authentication and message authentication [11,19,20,25]. In order to ensure the communication security. The communication entity should perform mutual authentication before communication. As long as the mutual authentication is implemented, some known attacks can be excluded.
- Integrity: preventing the altering GPS coordinates or disseminating of false information [25], thus ensuring the consistent and uncompromising adherence of data message over their whole passage through the flying networks [11,19,20]
- Confidentiality: Only the authorized UAVs are allowed to access the data packets [11,13,19,20,25].

- Identity anonymity: The UAV communication control system should keep identity anonymity from the attacker to ensure the users real identity is not obtained from eavesdropped or captured messages [11].
- Availability: The UAV communication control system should be always available to provide all services in any time and in any conditions [11,25].
- Privacy: By tracking the messages sent out by the same UAV at different locations, adversaries can disclosure the UAVs' identities and perform further analysis to get other information from the UAVs [11,18,20].
- Non-repudiation: Repudiation threat comes from the UAVs denying their behaviors in the IoD. For example, malicious UAVs abuse their valid identities to broadcast fake information in the IoD [18,20,25].
- DoS attack: DoS attack means that a malicious node attempts to exhaust energy resources of UAVs or disturb the network and routing protocol [15,20,25].
- Spoofing attack: The attacker could generate a spoofed message such that the receiver gets the incorrect message [15,25].

3. The Proposed Scheme

This section includes nine subsections: (1) system architecture is designed and described in Section 3.1, (2) the used notations in this study are defined in Section 3.2, (3) the manufacturer (UAV) registration phase of the proposed scheme is illustrated in Section 3.3, (4) the player (mobile device) registration phase of the proposed scheme is presented in Section 3.4, (5) the ground control station registration phase of the proposed scheme is described in Section 3.5, (6) the player and manufacturer authentication and communication phase of the proposed scheme is shown in Section 3.6, (7) the player and ground control station authentication and communication phase of the proposed scheme is designed in Section 3.7, (8) the player, UAV, and ground control station authentication and communication phase of the proposed scheme is discussed in Section 3.8, and (9) the ground control station and UAV authentication and communication phase of the proposed scheme is illustrated in Section 3.9.

3.1. System Architecture

Figure 1 is the system framework of the proposed scheme in this study.

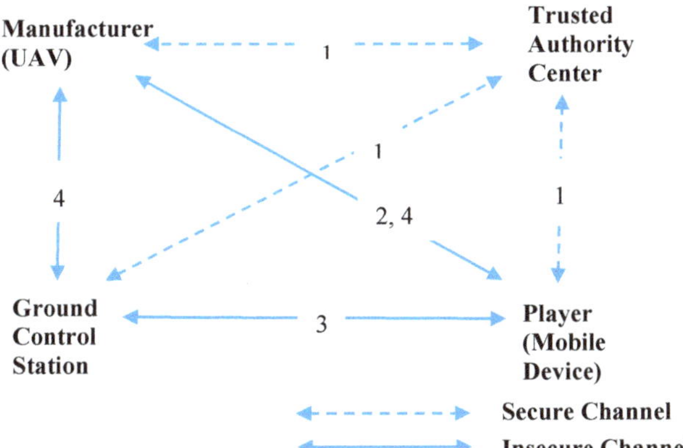

Figure 1. The framework of a traceable and privacy-preserving authentication for UAV ad hoc communication.

There are four parties in the scheme:

(1) Trusted authority center: a trusted third party agency which provides a public key and private key to the registrant.

(2) Manufacturer (UAV): a UAV manufacturing company. The company has jurisdiction over all manufactured UAVs.

(3) Player (mobile device): a person who intends to control a UAV. He/she must first buy or rent a UAV from the manufacturer, then obtain the flight permit before he/she can control the UAV.

(4) Ground control station (GCS): a control center that provides the facilities for human control of the UAV. A GCS reviews the flight path proposed by the player, and decides whether to agree to the flight request.

1. All UAVs manufactured, all mobile devices carried by players, and all ground control stations must be registered to the trusted authority center through a secure channel. The manufacturer (UAV), player (with mobile device), and ground control station sends their universally unique IDs to the trusted authority center. The trusted authority center returns parameters calculated by elliptic curve group technology.

2. When a player wants to control UAVs, the player carries his/her mobile device to buy or rent a UAV from the manufacturer. After mutual authentication between the player and the manufacturer, the manufacturer will transfer the purchase or rental certificate of the UAV to the player, and store the certificate to the UAV.

3. After the player has the right to use the UAV, then he/she must submit flight information and a purpose to the ground control station for review. After mutual authentication between the player and the ground control station, the ground control station will transfer the decision of the flight plan to the player, and keep the relevant flight information.

4. The player transfers the purchase or rental certificate of the UAV, and the flight path agreed by the ground control station to the UAV. After mutual authentication between the player and the UAV and mutual authentication between the UAV and the ground control station, the ground control station will confirm the legality of the UAV flight path. Once the legality of the relevant identity and flight path have been confirmed, the player can control the UAV through his/her mobile device.

3.2. Notations

q:	A k-bit prime
F_q:	A prime finite field
E/F_q:	An elliptic curve E over F_q
G:	A cyclic additive group of composite order q
P:	A generator for the group G
s:	A secret key of the trusted authority center
PK_{TAC}:	A public key of the trusted authority center, $PK_{TAC} = sP$
$H_i(\ )$:	ith one-way hash function
ID_x:	x's identity, like a universal unique ID code
r_x, a, b, c, d, e, f:	A random numbers of elliptic curve group
S_x:	x's elliptic curve group signature
SEK_{xy}:	A session key established by x and y
$E_x(m)$:	Use a session key x to encrypt the message m
$D_x(m)$:	Use a session key x to decrypt the message m
Sig_{xy}:	The signed message for parties x and y
SK_x/PK_x:	x's private key SK_x/x's public key PK_x
$S_{SK_x}(m)$:	Use x's private key SK_x to sign the message m

$V_{PK_x}(m)$:	Use x's public key PK_x to verify the message m
CHK_x:	x's verified message
$A \overset{?}{=} B$:	Determines if A is equal to B
$M_{payment}$:	The payment message between the player and the manufacturer (UAV)
$M_{request}$:	The flight plan proposed by the player
$M_{confirm}$:	The flight permission issued by ground control station to UAV
M_{GPS}:	The GPS message reported by the UAV
c_i:	The session key encrypted sensitive information
$Cert_{UAV}$:	The purchase or rental certificate of the UAV held by the player

3.3. Manufacturer (UAV) Registration Phase

The manufacturer must take the UAV to register with the trusted authority center. The manufacturer (UAV) registration phase of the proposed scheme is shown in Figure 2.

Step 1: The manufacturer selects an identity ID_{UAV}, and transmits it to the trusted authority center.

Step 2: The trusted authority center selects a random number r_{UAV}, calculates

$$R_{UAV} = r_{UAV}P,$$
$$h_{UAV} = H_1(ID_{UAV}, R_{UAV}),$$
$$S_{UAV} = r_{UAV} + h_{UAV}s,$$

and then sends $(R_{UAV}, S_{UAV}, PK_{UAV}, SK_{UAV})$ to the manufacturer.

Step 3: The manufacturer verifies

$$S_{UAV}P \overset{?}{=} R_{UAV} + H_1(ID_{UAV}, R_{UAV})PK_{TAC}.$$

If the verification is passed, the manufacturer stores $(R_{UAV}, S_{UAV}, PK_{UAV}, SK_{UAV})$ to the UAV.

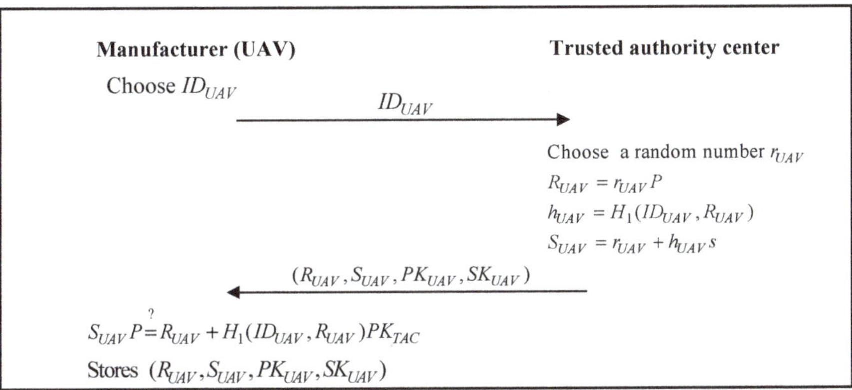

Figure 2. Manufacturer (UAV) registration phase of the proposed scheme.

3.4. Player (Mobile Device) Registration Phase

The player must take the mobile device to register with the trusted authority center. The scenarios of player (mobile device) registration phase is shown in Figure 3.

Step 1: The player selects an identity ID_{PMD}, and transmits it to the trusted authority center.

Step 2: The trusted authority center selects a random number r_{PMD}, calculates

$$R_{PMD} = r_{PMD}P,$$
$$h_{PMD} = H_1(ID_{PMD}, R_{PMD}),$$
$$S_{PMD} = r_{PMD} + h_{PMD}s,$$

and then sends $(R_{PMD}, S_{PMD}, PK_{PMD}, SK_{PMD})$ to the player.

Step 3: The player verifies

$$S_{PMD}P \stackrel{?}{=} R_{PMD} + H_1(ID_{PMD}, R_{PMD})PK_{TAC}.$$

If the verification is passed, the player stores $(R_{PMD}, S_{PMD}, PK_{PMD}, SK_{PMD})$ to the mobile device.

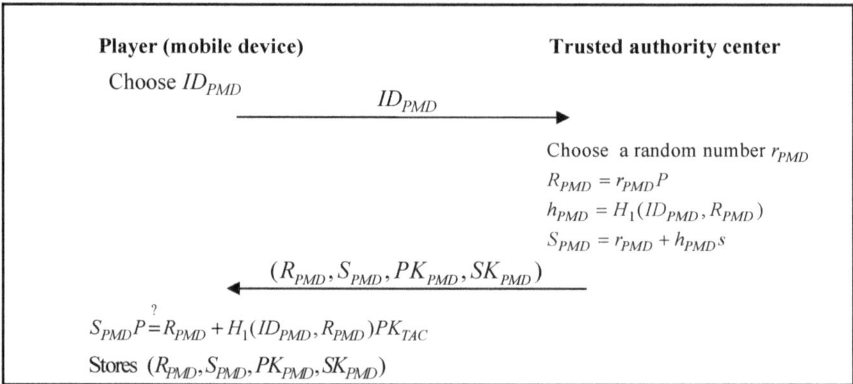

Figure 3. Player (mobile device) registration phase of the proposed scheme.

3.5. Ground Control Station Registration Phase

The ground control station must also register with the trusted authority center. The ground control station registration phase of the proposed scheme is shown in Figure 4.

Step 1: The ground control station selects an identity ID_{GCS}, and transmits it to the trusted authority center.

Step 2: The trusted authority center selects a random number r_{GCS}, calculates

$$R_{GCS} = r_{GCS}P,$$
$$h_{GCS} = H_1(ID_{GCS}, R_{GCS}),$$
$$S_{GCS} = r_{GCS} + h_{GCS}s,$$

and then sends $(R_{GCS}, S_{GCS}, PK_{GCS}, SK_{GCS})$ to the ground control station.

Step 3: The ground control station verifies

$$S_{GCS}P \stackrel{?}{=} R_{GCS} + H_1(ID_{GCS}, R_{GCS})PK_{TAC}.$$

If the verification is passed, the ground control station stores $(R_{GCS}, S_{GCS}, PK_{GCS}, SK_{GCS})$.

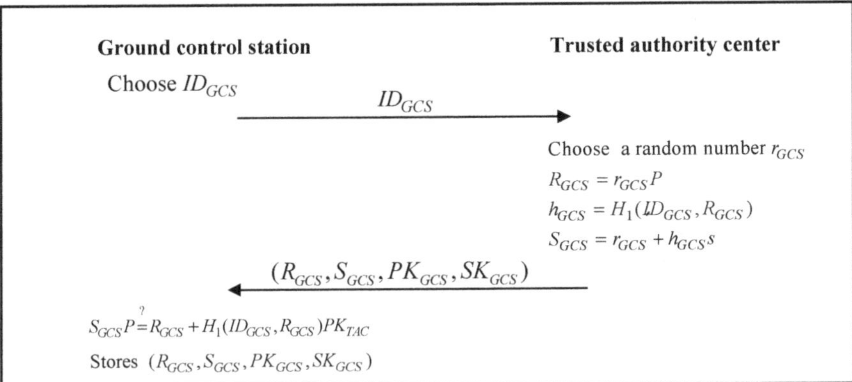

Figure 4. Ground control station registration phase of the proposed scheme.

3.6. Player and Manufacturer Authentication and Communication Phase

When a player wants to control UAVs, the player carries his/her mobile device to buy or rent a UAV from the manufacturer. After mutual authentication between the player and the manufacturer, the manufacturer will transfer the purchase or rental certificate of the UAV to the player, and store the certificate of the UAV. The player and manufacturer authentication and communication phase is shown in Figure 5.

Step 1: The player selects a random number a, computes

$$T_{PMD} = aP,$$

and then transmits $(ID_{PMD}, R_{PMD}, T_{PMD})$ to the manufacturer.

Step 2: The manufacturer selects a random number b, calculates

$$T_{UAV} = bP,$$
$$PK_{PMD} = R_{PMD} + H_1(ID_{PMD}, R_{PMD})PK_{TAC},$$
$$K_{UP1} = S_{UAV}T_{PMD} + bPK_{PMD},$$
$$K_{UP2} = bT_{PMD},$$

and the session key

$$SEK_{UP} = H_2(K_{UP1}, K_{UP2}).$$

The manufacturer then calculates

$$CHK_{PU} = H_3(SEK_{UP}, T_{PMD})$$

and transmits $(ID_{UAV}, R_{UAV}, T_{UAV}, CHK_{PU})$ to the player.

Step 3: The player calculates

$$PK_{UAV} = R_{UAV} + H_1(ID_{UAV}, R_{UAV})PK_{TAC},$$
$$K_{PU1} = S_{PMD}T_{UAV} + aPK_{UAV},$$
$$K_{PU2} = aT_{UAV},$$

and the session key

$$SEK_{UP} = H_2(K_{PU1}, K_{PU2}),$$

The player verifies

$$CHK_{PU} \overset{?}{=} H_3(SEK_{UP}, T_{PMD})$$

to check the legality of the manufacturer. If the verification is passed, the player computes

$$c_{PMD} = E_{SEK_{UP}}(M_{payment}),$$
$$CHK_{UP} = H_3(SEK_{UP}, T_{UAV}),$$

and transmits $(ID_{PMD}, c_{PMD}, CHK_{UP})$ to the manufacturer.

Step 4: The manufacturer verifies

$$CHK_{UP} \overset{?}{=} H_3(SEK_{UP}, T_{UAV})$$

to check the legality of the player. If the verification is passed, the session key SEK_{UP} between the player and the manufacturer is established successfully. The manufacturer calculates

$$M_{payment} = D_{SEK_{UP}}(c_{PMD})$$

to get the payment information of the player. After the payment, the manufacturer generates the encrypted purchase or rental certificate of the UAV

$$c_{UAV} = E_{SEK_{UP}}(M_{payment}, Cert_{UAV}),$$
$$Sig_{UAV} = S_{SK_{UAV}}(M_{payment}, Cert_{UAV}),$$

and transmits $(ID_{UAV}, c_{UAV}, Sig_{UAV})$ to the player.

Step 5: The player decrypts the received message

$$(M_{payment}, Cert_{UAV}) = D_{SEK_{UP}}(c_{UAV}),$$

verifies the signature

$$(M_{payment}, Cert_{UAV}) \overset{?}{=} V_{PK_{UAV}}(Sig_{UAV}),$$

and obtains the purchase or rental certificate of the UAV from the manufacturer.

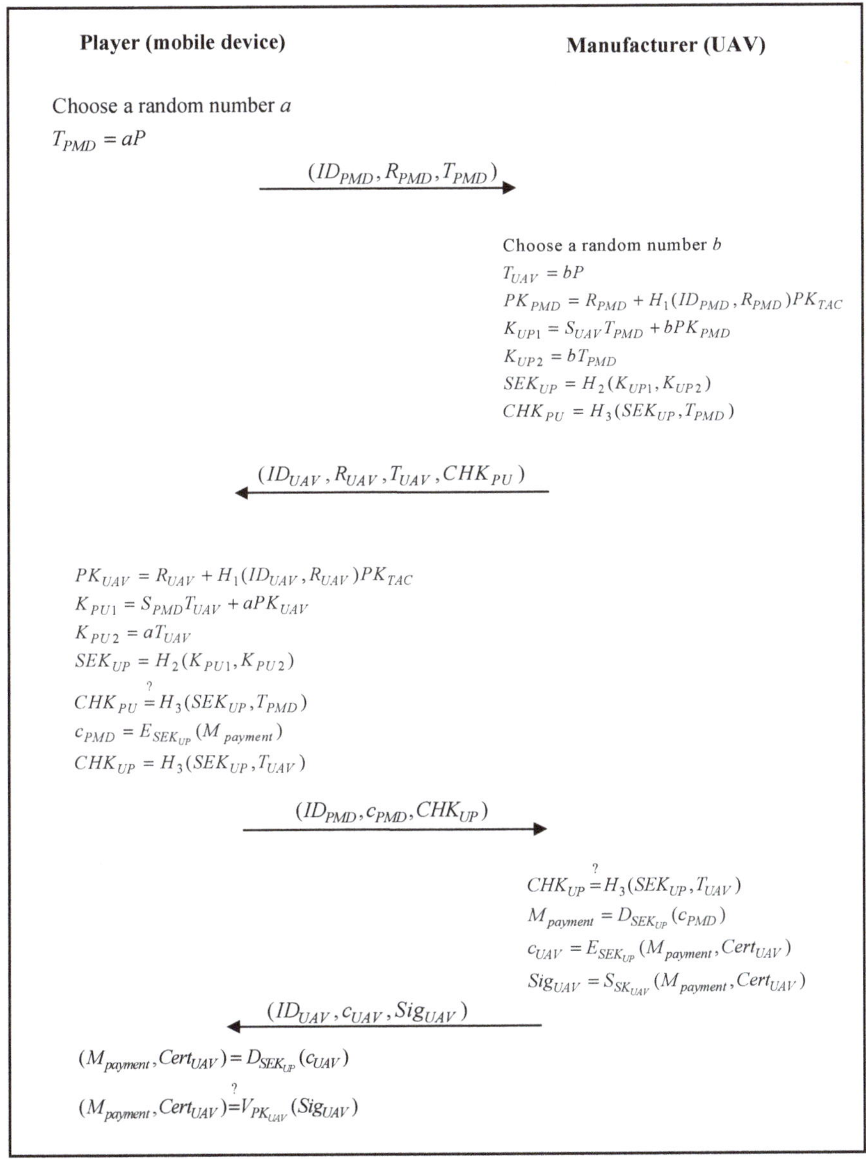

Figure 5. Player and manufacturer authentication and communication phase of the proposed scheme.

3.7. Player and Ground Control Station Authentication and Communication Phase

After the player has the right to use the UAV, then he/she must submit a flight path and purpose to the ground control station for review. After mutual authentication between the player and the ground control station, the ground control station will transfer the decision of the flight plan to the player, and keeps the relevant flight information. The player and ground control station authentication and communication phase of the proposed scheme is shown in Figure 6.

Step 1: The player selects a random number c, computes

$$T_{PMD2} = cP,$$

and then transmits $(ID_{PMD}, R_{PMD}, T_{PMD2})$ to the ground control station.

Step 2: The ground control station selects a random number d, calculates

$$T_{GCS} = dP,$$
$$PK_{PMD} = R_{PMD} + H_1(ID_{PMD}, R_{PMD})PK_{TAC},$$
$$K_{GP1} = S_{GCS}T_{PMD2} + dPK_{PMD},$$
$$K_{GP2} = dT_{PMD2},$$

and the session key

$$SEK_{GP} = H_2(K_{GP1}, K_{GP2}).$$

The ground control station then calculates

$$CHK_{PG} = H_3(SEK_{GP}, T_{PMD2})$$

and transmits $(ID_{GCS}, R_{GCS}, T_{GCS}, CHK_{PG})$ to the player.

Step 3: The player calculates

$$PK_{GCS} = R_{GCS} + H_1(ID_{GCS}, R_{GCS})PK_{TAC},$$
$$K_{PG1} = S_{PMD}T_{GCS} + cPK_{GCS},$$
$$K_{PG2} = cT_{GCS},$$

and the session key

$$SEK_{GP} = H_2(K_{PG1}, K_{PG2}).$$

The player verifies

$$CHK_{PG} \stackrel{?}{=} H_3(SEK_{GP}, T_{PMD2})$$

to check the legality of the ground control station. If the verification is passed, the player calculates

$$c_{PMD2} = E_{SEK_{GP}}(M_{request}, Cert_{UAV}),$$
$$CHK_{GP} = H_3(SEK_{GP}, T_{GCS}),$$

and transmits $(ID_{PMD}, c_{PMD2}, CHK_{GP})$ to the ground control station.

Step 4: The ground control station verifies

$$CHK_{GP} \stackrel{?}{=} H_3(SEK_{GP}, T_{GCS})$$

to check the legality of the player. If the verification is passed, the session key SEK_{GP} between the player and the ground control station is established successfully. The ground control station calculates

$$(M_{request}, Cert_{UAV}) = D_{SEK_{GP}}(c_{PMD2})$$

to get the flight path information of the player. After the review, the ground control station generates the encrypted decision of the flight plan

$$c_{GCS} = E_{SEK_{GP}}(ID_{PMD}, M_{request}, Cert_{UAV}),$$
$$Sig_{GCS} = S_{SK_{GCS}}(ID_{PMD}, M_{request}, Cert_{UAV}),$$

and transmits $(ID_{GCS}, c_{GCS}, Sig_{GCS})$ to the player.

Step 5: The player decrypts the received message

$$(ID_{PMD}, M_{request}, Cert_{UAV}) = D_{SEK_{GP}}(c_{GCS}),$$

verifies the signature

$$(ID_{PMD}, M_{request}, Cert_{UAV}) \overset{?}{=} V_{PK_{GCS}}(Sig_{GCS}),$$

and obtains the decision of the flight plan from the ground control station.

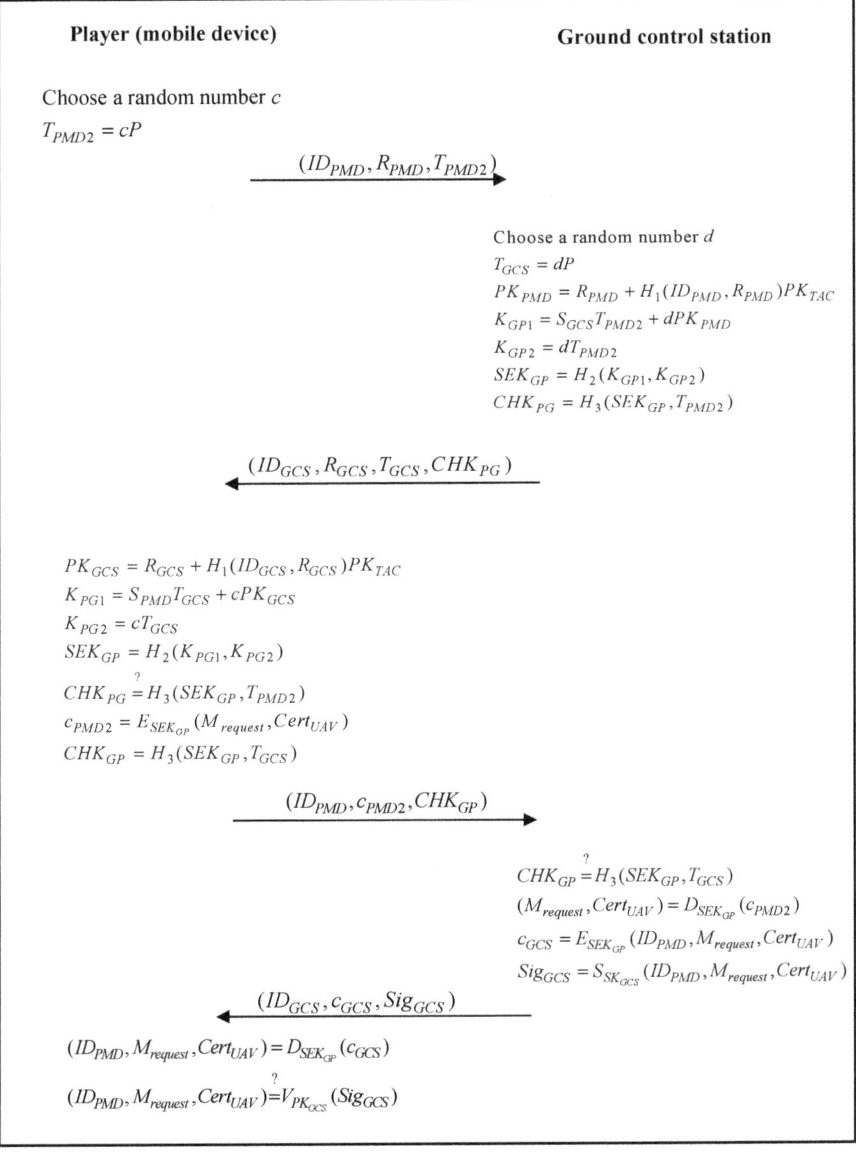

Figure 6. Player and ground control station authentication and communication phase of the proposed scheme.

3.8. Player, UAV and Ground Control Station Authentication and Communication Phase

The player transfers the purchase or rental certificate of the UAV, and the flight path agreed by the ground control station to the UAV. After mutual authentication between the player and the UAV, and mutual authentication between the UAV and the ground control station, the UAV will confirm the legality of the flight path again from the ground control station. After confirming the legality of the relevant identity and flight path, the player can control the UAV through his/her mobile device. The player, UAV and ground control station authentication and communication phase of the proposed scheme is shown in Figure 7.

Step 1: The player calculates

$$c_{PMD3} = E_{SEK_{UP}}(M_{request}, Cert_{UAV}),$$
$$Sig_{PMD3} = S_{SK_{PMD}}(M_{request}, Cert_{UAV}),$$

and transmits $(ID_{PMD}, c_{PMD3}, Sig_{PMD3})$ to the UAV.

Step 2: The UAV decrypts the received message

$$(M_{request}, Cert_{UAV}) = D_{SEK_{UP}}(c_{PMD3}),$$

verifies the signature

$$(M_{request}, Cert_{UAV}) \overset{?}{=} V_{PK_{PMD}}(Sig_{PMD3}),$$

and obtains the purchase or rental certificate of the UAV, and the flight path agreed by the ground control station.

The UAV then chooses a random number e, calculates

$$T_{UAV2} = eP,$$

and then transmits $(ID_{UAV}, R_{UAV}, T_{UAV2})$ to the ground control station.

Step 3: The ground control station chooses a random number f, computes

$$T_{GCS2} = fP,$$
$$PK_{UAV} = R_{UAV} + H_1(ID_{UAV}, R_{UAV})PK_{TAC},$$
$$K_{GU1} = S_{GCS}T_{UAV2} + fPK_{UAV},$$
$$K_{GU2} = fT_{UAV2},$$

and the session key

$$SEK_{GU} = H_2(K_{GU1}, K_{GU2}).$$

The ground control station then calculates

$$CHK_{UG} = H_3(SEK_{GU}, T_{UAV2}),$$

and transmits $(ID_{GCS}, R_{GCS}, T_{GCS2}, CHK_{UG})$ to the UAV.

Step 4: The UAV calculates

$$PK_{GCS} = R_{GCS} + H_1(ID_{GCS}, R_{GCS})PK_{TAC},$$
$$K_{UG1} = S_{UAV}T_{GCS2} + ePK_{GCS},$$
$$K_{UG2} = eT_{GCS2},$$

and the session key

$$SEK_{GU} = H_2(K_{UG1}, K_{UG2}).$$

The UAV verifies

$$CHK_{UG} \overset{?}{=} H_3(SEK_{GU}, T_{UAV2})$$

to check the legality of the ground control station. If the verification is passed, the UAV calculates

$$c_{UAV2} = E_{SEK_{GU}}(ID_{PMD}, M_{request}, Cert_{UAV}),$$
$$CHK_{GU} = H_3(SEK_{GU}, T_{GCS2}),$$

and transmits $(ID_{UAV}, c_{UAV2}, CHK_{GU})$ to the ground control station.

Step 5: The ground control station verifies

$$CHK_{UG} \stackrel{?}{=} H_3(SEK_{GU}, T_{GCS2})$$

to check the legality of the UAV. If the verification is passed, the session key SEK_{GU} between the UAV and the ground control station is established successfully. The ground control station calculates

$$(ID_{PMD}, M_{request}, Cert_{UAV}) = D_{SEK_{GU}}(c_{UAV2})$$

to get the flight path information of the UAV. After the review, the ground control station generates the encrypted confirm message of the flight plan

$$c_{GCS2} = E_{SEK_{GU}}(ID_{PMD}, M_{confirm}, Cert_{UAV}),$$
$$Sig_{GCS2} = S_{SK_{GCS}}(ID_{PMD}, M_{confirm}, Cert_{UAV}),$$

and transmits $(ID_{GCS}, c_{GCS2}, Sig_{GCS2})$ to the UAV.

Step 6: The UAV decrypts the received message

$$(ID_{PMD}, M_{confirm}, Cert_{UAV}) = D_{SEK_{GU}}(c_{GCS2}),$$

verifies the signature

$$(ID_{PMD}, M_{confirm}, Cert_{UAV}) \stackrel{?}{=} V_{PK_{GCS}}(Sig_{GCS2}),$$

and obtains the confirm message of the flight plan from the ground control station. Then, the UAV generates the encrypted confirm message of the flight plan and GPS information

$$c_{UAV3} = E_{SEK_{UP}}(ID_{PMD}, M_{confirm}, M_{GPS}, Cert_{UAV}),$$
$$Sig_{UAV3} = S_{SK_{UAV}}(ID_{PMD}, M_{confirm}, M_{GPS}, Cert_{UAV}),$$

and transmits $(ID_{UAV}, c_{UAV3}, Sig_{UAV3})$ to the player.

Step 7: The player decrypts the received message

$$(ID_{PMD}, M_{request}, M_{GPS}, Cert_{UAV}) = D_{SEK_{UP}}(c_{UAV3}),$$

verifies the signature

$$(ID_{PMD}, M_{confirm}, M_{GPS}, Cert_{UAV}) \stackrel{?}{=} V_{PK_{UAV}}(Sig_{UAV3}),$$

then obtains the confirm message of the flight plan and GPS information.

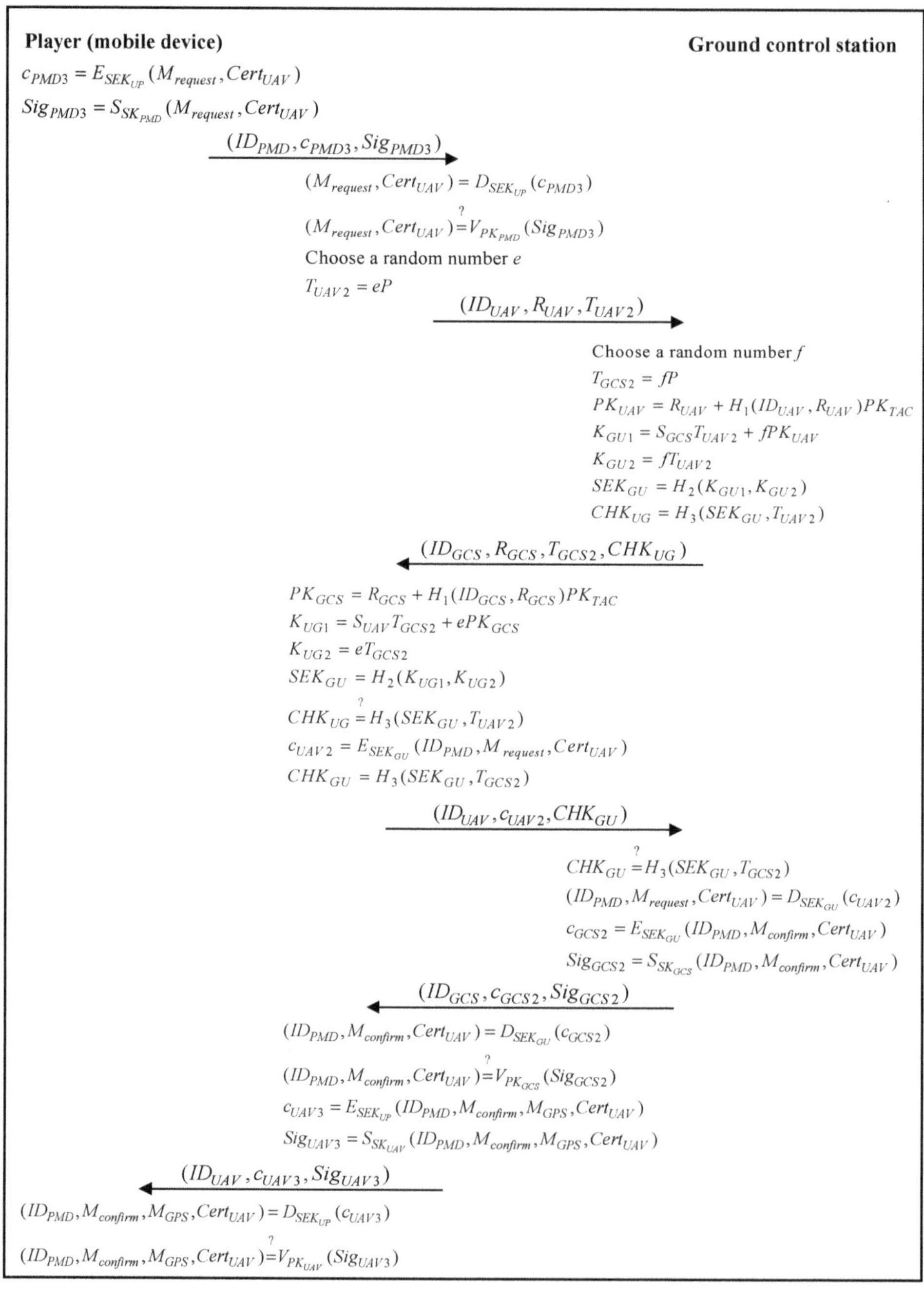

Figure 7. Player, UAV, and ground control station authentication and communication phase of the proposed scheme.

3.9. Ground Control Station and UAV Authentication and Communication Phase

When the ground control station wants to know whether the scope of the regulation has been applied to the UAV, the ground control station can ask the UAV to provide relevant proof. After mutual authentication between the ground control station and the UAV, the UAV will respond and confirm the message of the flight plan from the ground control station and GPS information to the ground control station. The ground control station and UAV authentication and communication phase of the proposed scheme is shown in Figure 8.

Step 1: The ground control station calculates

$$c_{GCS3} = E_{SEK_{GU}}(ID_{UAV}, M_{request}),$$
$$Sig_{GCS3} = S_{SK_{GCS}}(ID_{UAV}, M_{request}),$$

and transmits $(ID_{UAV}, M_{request}) = D_{SEK_{GU}}(c_{GCS3})$ to the UAV.

Step 2: The UAV decrypts the received message

$$(ID_{UAV}, M_{request}) = D_{SEK_{GU}}(c_{GCS3}),$$

verifies the signature

$$(ID_{UAV}, M_{request}) \overset{?}{=} V_{PK_{GCS}}(Sig_{GCS3}),$$

and obtains the legality check request from the ground control station. Then, the UAV generates the encrypted confirmation message of the flight plan and GPS information

$$C_{UAV4} = E_{SEK_{GU}}(ID_{PMD}, M_{confirm}, M_{GPS}, Cert_{UAV}),$$
$$Sig_{UAV4} = S_{SK_{UAV}}(ID_{PMD}, M_{confirm}, M_{GPS}, Cert_{UAV}),$$

and transmits $(ID_{UAV}, c_{UAV4}, Sig_{UAV4})$ to the ground control station.

Step 3: The ground control station decrypts the received message

$$(ID_{PMD}, M_{confirm}, M_{GPS}, Cert_{UAV}) = D_{SEK_{GU}}(c_{UAV4}),$$

verifies the signature

$$(ID_{PMD}, M_{confirm}, M_{GPS}, Cert_{UAV}) \overset{?}{=} V_{PK_{UAV}}(Sig_{UAV4}),$$

then obtains the response of the UAV and GPS information.

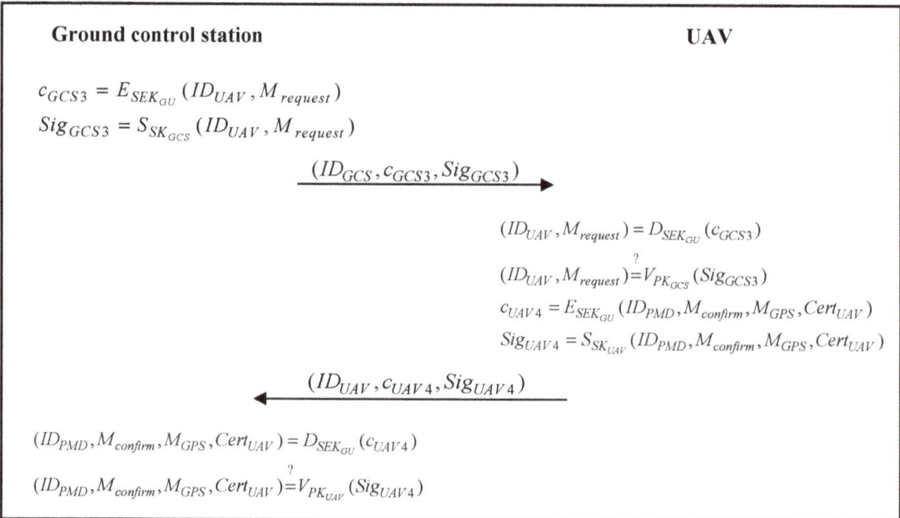

Figure 8. Ground control station and UAV authentication and communication phase of the proposed scheme.

4. Security Analysis

This section includes nine subsections: (1) the mutual authentication of the proposed scheme is analyzed in Section 4.1, (2) the integrity and confidentiality of the proposed scheme are evaluated in Section 4.2, (3) the identity anonymity and privacy of the proposed scheme are proved in Section 4.3, (4) availability and prevention of DoS attack are discussed in Section 4.4, (5) prevention of spoofing attack is discussed in Section 4.5, (6) the non-repudiation of the proposed scheme is analyzed in Section 4.6, (7) security issues are compared in Section 4.7, (8) the computation cost of the proposed scheme is compared with other schemes in Section 4.8, and (9) the communication cost of the proposed scheme is compared with other schemes in Section 4.9.

4.1. Mutual Authentication

BAN logic [26] is used to prove that the proposed scheme achieves mutual authentication between different parties in each phase.

In the player and manufacturer authentication and communication phase, the main goal of the scheme is to make sure whether the legality is authenticated by the player P and the manufacturer M.

$$G1: \quad P| \equiv P \stackrel{SEK_{UP}}{\leftrightarrow} M$$

$$G2: \quad P| \equiv M| \equiv P \stackrel{SEK_{UP}}{\leftrightarrow} M$$

$$G3: \quad M| \equiv P \stackrel{SEK_{UP}}{\leftrightarrow} M$$

$$G4: \quad M| \equiv P| \equiv P \stackrel{SEK_{UP}}{\leftrightarrow} M$$

$$G5: \quad P| \equiv ID_{UAV}$$

$$G6: \quad P| \equiv M| \equiv ID_{UAV}$$

$$G7: \quad M| \equiv ID_{PMD}$$

$$G8: \quad M| \equiv P| \equiv ID_{PMD}$$

According to the player and manufacturer authentication and communication phase, BAN logic is used to produce an idealized form as follows.

$$M1: \ (< ID_{PMD}, R_{PMD}, T_{PMD} >_{PK_{UAV}}, < H(SEK_{UP}, T_{UAV}) >_{CHK_{UP}})$$
$$M2: \ (< ID_{UAV}, R_{UAV}, T_{UAV} >_{PK_{PMD}}, < H(SEK_{UP}, T_{PMD}) >_{CHK_{PU}})$$

To analyze the proposed scheme, the following assumptions are made.

$$A1: \ P| \equiv \#(T_{PMD})$$
$$A2: \ M| \equiv \#(T_{PMD})$$
$$A3: \ P| \equiv \#(T_{UAV})$$
$$A4: \ M| \equiv \#(T_{UAV})$$
$$A5: \ P| \equiv M| \Rightarrow P \overset{SEK_{UP}}{\leftrightarrow} M$$
$$A6: \ M| \equiv P| \Rightarrow P \overset{SEK_{UP}}{\leftrightarrow} M$$
$$A7: \ P| \equiv M| \Rightarrow ID_{UAV}$$
$$A8: \ M| \equiv P| \Rightarrow ID_{PMD}$$

According to these assumptions and goals of BAN logic, the main proof of the player and manufacturer authentication and communication phase is as follows.

a.   The manufacturer M authenticates the player P.

By *M1* and the *seeing rule, Statement 1* can be derived.

$$M \triangleleft (< ID_{PMD}, R_{PMD}, T_{PMD} >_{PK_{UAV}}, < H(SEK_{UP}, T_{UAV}) >_{CHK_{UP}}). \tag{Statement 1}$$

By *A2* and the *freshness rule, Statement 2* can be derived.

$$M| \equiv \#(< ID_{PMD}, R_{PMD}, T_{PMD} >_{PK_{UAV}}, < H(SEK_{UP}, T_{UAV}) >_{CHK_{UP}}). \tag{Statement 2}$$

By (*Statement 1*), *A4*, and the *message meaning rule, Statement 3* can be derived.

$$M| \equiv P| \sim (< ID_{PMD}, R_{PMD}, T_{PMD} >_{PK_{UAV}}, < H(SEK_{UP}, T_{UAV}) >_{CHK_{UP}}). \tag{Statement 3}$$

By (*Statement 2*), (*Statement 3*), and the *nonce verification rule, Statement 4* can be derived.

$$M| \equiv \#(< ID_{PMD}, R_{PMD}, T_{PMD} >_{PK_{UAV}}, < H(SEK_{UP}, T_{UAV}) >_{CHK_{UP}}). \tag{Statement 4}$$

By (*Statement 4*) and the *belief rule, Statement 5* can be derived.

$$M| \equiv P| \equiv P \overset{SEK_{UP}}{\leftrightarrow} M. \tag{Statement 5}$$

By (*Statement 5*), *A6*, and the *jurisdiction rule, Statement 6* can be derived.

$$M| \equiv P \overset{SEK_{UP}}{\leftrightarrow} M. \tag{Statement 6}$$

By (*Statement 6*) and the *belief rule, Statement 7* can be derived.

$$M| \equiv P| \equiv ID_{PMD}. \tag{Statement 7}$$

By (*Statement 7*), *A8*, and the *jurisdiction rule, Statement 8* can be derived.

$$M| \equiv ID_{PMD}. \tag{Statement 8}$$

b.   The player P authenticates the manufacturer M.

By *M2* and the *seeing rule, Statement 9* can be derived.

$$P \triangleleft (< ID_{UAV}, R_{UAV}, T_{UAV} >_{PK_{PMD}}, < H(SEK_{UP}, T_{PMD}) >_{CHK_{PU}}). \tag{Statement 9}$$

By *A1* and the *freshness rule, Statement 10* can be derived.

$$P| \equiv \#(< ID_{UAV}, R_{UAV}, T_{UAV} >_{PK_{PMD}}, < H(SEK_{UP}, T_{PMD}) >_{CHK_{PU}}). \tag{Statement 10}$$

By (*Statement 9*), *A3*, and the *message meaning rule, Statement 11* can be derived.

$$P| \equiv M| \sim (< ID_{UAV}, R_{UAV}, T_{UAV} >_{PK_{PMD}}, < H(SEK_{UP}, T_{PMD}) >_{CHK_{PU}}). \tag{Statement 11}$$

By (*Statement 10*), (*Statement 11*), and the *nonce verification rule*, Statement 12 can be derived.

$$P| \equiv M| \equiv (< ID_{UAV}, R_{UAV}, T_{UAV} >_{PK_{PMD}}, < H(SEK_{UP}, T_{PMD}) >_{CHK_{Pu}}). \tag{Statement 12}$$

By (*Statement 12*) and the *belief rule*, Statement 13 can be derived.

$$P| \equiv M| \equiv P \overset{SEK_{UP}}{\leftrightarrow} M. \tag{Statement 13}$$

By (*Statement 13*), A5, and the *jurisdiction rule*, Statement 14 can be derived.

$$P| \equiv P \overset{SEK_{UP}}{\leftrightarrow} M. \tag{Statement 14}$$

By (*Statement 14*) and the *belief rule*, Statement 15 can be derived.

$$P| \equiv M| \equiv ID_{UAV}. \tag{Statement 15}$$

By (*Statement 15*), A7, and the *jurisdiction rule*, Statement 16 can be derived.

$$P| \equiv ID_{UAV}. \tag{Statement 16}$$

By (*Statement 6*), (*Statement 8*), (*Statement 14*), and (*Statement 16*), it can be proved that the player P and the manufacturer M authenticate each other in the proposed scheme. Moreover, it can also be proved that the proposed scheme can establish a session key between the player P and the manufacturer M.

In the proposed scheme, the manufacturer authenticates the player by

$$CHK_{UP} \overset{?}{=} H_3(SEK_{UP}, T_{UAV}).$$

If it passes the verification, the manufacturer authenticates the legality of the player. The player authenticates the manufacturer by

$$CHK_{PU} \overset{?}{=} H_3(SEK_{UP}, T_{PMD}).$$

If it passes the verification, the player authenticates the legality of the manufacturer. The player and manufacturer authentication and communication phase of the proposed scheme thus guarantees mutual authentication between the player and the manufacturer.

In the player and ground control station authentication and communication phase, the main goal of the scheme is to make sure whether the legality is authenticated by the player P and the ground control station G.

$$G9: \ P| \equiv P \overset{SEK_{GP}}{\leftrightarrow} G$$

$$G10: P| \equiv G| \equiv P \overset{SEK_{GP}}{\leftrightarrow} G$$

$$G11: G| \equiv P \overset{SEK_{GP}}{\leftrightarrow} G$$

$$G12: G| \equiv P| \equiv P \overset{SEK_{GP}}{\leftrightarrow} G$$

$$G13: P| \equiv ID_{GCS}$$

$$G14: P| \equiv G| \equiv ID_{GCS}$$

$$G15: G| \equiv ID_{PMD}$$

$$G16: G| \equiv P| \equiv ID_{PMD}$$

According to the player and ground control station authentication and communication phase, BAN logic is used to produce an idealized form as follows.

$$M3: \ (< ID_{PMD}, R_{PMD}, T_{PMD2} >_{PK_{GCS}}, < H(SEK_{GP}, T_{GCS}) >_{CHK_{GP}})$$

$$M4: \ (< ID_{GCS}, R_{GCS}, T_{GCS} >_{PK_{PMD}}, < H(SEK_{GP}, T_{PMD2}) >_{CHK_{PG}})$$

To analyze the proposed scheme, the following assumptions are made.

$$A9: \ P| \equiv \#(T_{PMD2})$$

$$A10: G| \equiv \#(T_{PMD2})$$

$$A11: P| \equiv \#(T_{GCS})$$

$A12: G| \equiv \#(T_{GCS})$

$A13: P| \equiv G| \Rightarrow P \overset{SEK_{GP}}{\leftrightarrow} G$

$A14: G| \equiv P| \Rightarrow P \overset{SEK_{GP}}{\leftrightarrow} G$

$A15: P| \equiv G| \Rightarrow ID_{GCS}$

$A16: G| \equiv P| \Rightarrow ID_{PMD}$

According to these assumptions and goals of BAN logic, the main proof of the player and ground control station authentication and communication phase is as follows.

c. The ground control station G authenticates the player P.

By *M3* and the *seeing rule*, *Statement 17* can be derived.

$$G \triangleleft (< ID_{PMD}, R_{PMD}, T_{PMD2} >_{PK_{GCS}}, < H(SEK_{GP}, T_{GCS}) >_{CHK_{GP}}).$$ *(Statement 17)*

By *A10* and the *freshness rule*, *Statement 18* can be derived.

$$G| \equiv \#(< ID_{PMD}, R_{PMD}, T_{PMD2} >_{PK_{GCS}}, < H(SEK_{GP}, T_{GCS}) >_{CHK_{GP}}).$$ *(Statement 18)*

By *(Statement 17)*, *A12*, and the *message meaning rule*, *Statement 19* can be derived.

$$G| \equiv P| \sim (< ID_{PMD}, R_{PMD}, T_{PMD2} >_{PK_{GCS}}, < H(SEK_{GP}, T_{GCS}) >_{CHK_{GP}}).$$ *(Statement 19)*

By *(Statement 18)*, *(Statement 19)*, and the *nonce verification rule*, *Statement 20* can be derived.

$$G| \equiv P| \equiv (< ID_{PMD}, R_{PMD}, T_{PMD2} >_{PK_{GCS}}, < H(SEK_{GP}, T_{GCS}) >_{CHK_{GP}}).$$ *(Statement 20)*

By *(Statement 20)* and the *belief rule*, *Statement 21* can be derived.

$$G| \equiv P| \equiv P \overset{SEK_{GP}}{\leftrightarrow} G.$$ *(Statement 21)*

By *(Statement 21)*, *A14*, and the *jurisdiction rule*, *Statement 22* can be derived.

$$G| \equiv P \overset{SEK_{GP}}{\leftrightarrow} G.$$ *(Statement 22)*

By *(Statement 22)* and the *belief rule*, *Statement 23* can be derived.

$$G| \equiv P| \equiv ID_{PMD}.$$ *(Statement 23)*

By *(Statement 23)*, *A16*, and the *jurisdiction rule*, *Statement 24* can be derived.

$$G| \equiv ID_{PMD}.$$ *(Statement 24)*

d. The player P authenticates the ground control station G.

By *M4* and the *seeing rule*, *Statement 25* can be derived.

$$P \triangleleft (< ID_{GCS}, R_{GCS}, T_{GCS} >_{PK_{PMD}}, < H(SEK_{GP}, T_{PMD2}) >_{CHK_{PG}}).$$ *(Statement 25)*

By *A9* and the *freshness rule*, *Statement 26* can be derived.

$$P| \equiv \#(< ID_{GCS}, R_{GCS}, T_{GCS} >_{PK_{PMD}}, < H(SEK_{GP}, T_{PMD2}) >_{CHK_{PG}}).$$ *(Statement 26)*

By *(Statement 25)*, *A11*, and the *message meaning rule*, *Statement 27* can be derived.

$$P| \equiv G| \sim (< ID_{GCS}, R_{GCS}, T_{GCS} >_{PK_{PMD}}, < H(SEK_{GP}, T_{PMD2}) >_{CHK_{PG}}).$$ *(Statement 27)*

By *(Statement 26)*, *(Statement 27)*, and the *nonce verification rule*, *Statement 28* can be derived.

$$P| \equiv G| \equiv (< ID_{GCS}, R_{GCS}, T_{GCS} >_{PK_{PMD}}, < H(SEK_{GP}, T_{PMD2}) >_{CHK_{PG}}).$$ *(Statement 28)*

By *(Statement 28)* and the *belief rule*, *Statement 29* can be derived.

$$P| \equiv G| \equiv P \overset{SEK_{GP}}{\leftrightarrow} G.$$ *(Statement 29)*

By *(Statement 29)*, *A13*, and the *jurisdiction rule*, *Statement 30* can be derived.

$$P| \equiv P \overset{SEK_{GP}}{\leftrightarrow} G.$$ *(Statement 30)*

By *(Statement 30)* and the *belief rule*, *Statement 31* can be derived.

$$P| \equiv G| \equiv ID_{GCS}.$$ *(Statement 31)*

By *(Statement 31)*, *A15*, and the *jurisdiction rule*, *Statement 32* can be derived.

$$P| \equiv ID_{GCS}.$$ *(Statement 32)*

By *(Statement 22)*, *(Statement 24)*, *(Statement 30)*, and *(Statement 32)*, it can be proved that the player P and the ground control station G authenticate each other in the proposed scheme. Moreover, it can also be proved that the proposed scheme can establish a session key between the player P and the ground control station G.

In the proposed scheme, the ground control station authenticates the player by

$$CHK_{GP} \stackrel{?}{=} H_3(SEK_{GP}, T_{GCS}).$$

If it passes the verification, the manufacturer authenticates the legality of the player. The player authenticates the ground control station by

$$CHK_{PG} \stackrel{?}{=} H_3(SEK_{GP}, T_{PMD2}).$$

If it passes the verification, the player authenticates the legality of the ground control station. The player and ground control station authentication and communication phase of the proposed scheme thus guarantees mutual authentication between the player and the ground control station.

In the player, UAV, and ground control station authentication and communication phase, the main goal of the scheme is to make sure whether the legality is authenticated by the UAV U and the ground control station G.

$G17: U| \equiv U \stackrel{SEK_{GU}}{\leftrightarrow} G$

$G18: U| \equiv G| \equiv U \stackrel{SEK_{GU}}{\leftrightarrow} G$

$G19: G| \equiv U \stackrel{SEK_{GU}}{\leftrightarrow} G$

$G20: G| \equiv U| \equiv U \stackrel{SEK_{GU}}{\leftrightarrow} G$

$G21: U| \equiv ID_{GCS}$

$G22: U| \equiv G| \equiv ID_{GCS}$

$G23: G| \equiv ID_{UAV}$

$G24: G| \equiv U| \equiv ID_{UAV}$

According to the player, UAV, and ground control station authentication and communication phase, BAN logic is used to produce an idealized form as follows:

$M5: \ (< ID_{UAV}, R_{UAV}, T_{UAV2} >_{PK_{GCS}}, < H(SEK_{GU}, T_{GCS2}) >_{CHK_{GU}})$

$M6: \ (< ID_{GCS}, R_{GCS}, T_{GCS2} >_{PK_{UAV}}, < H(SEK_{GU}, T_{UAV2}) >_{CHK_{UG}})$

To analyze the proposed scheme, the following assumptions are made.

$A17: U| \equiv \#(T_{UAV2})$

$A18: G| \equiv \#(T_{UAV2})$

$A19: U| \equiv \#(T_{GCS2})$

$A20: G| \equiv \#(T_{GCS2})$

$A21: U| \equiv G| \Rightarrow U \stackrel{SEK_{GU}}{\leftrightarrow} G$

$A22: G| \equiv U| \Rightarrow U \stackrel{SEK_{GU}}{\leftrightarrow} G$

$A23: U| \equiv G| \Rightarrow ID_{GCS}$

$A24: G| \equiv U| \Rightarrow ID_{UAV}$

According to these assumptions and goals of BAN logic, the main proof of the player, UAV, and ground control station authentication and communication phase is as follows.

e   The ground control station G authenticates the UAV U.

By *M5* and the *seeing rule*, *Statement 33* can be derived.

$$G \lhd (< ID_{UAV}, R_{UAV}, T_{UAV2} >_{PK_{GCS}}, < \qquad \text{(Statement 33)}$$
$$H(SEK_{GU}, T_{GCS2}) >_{CHK_{Gu}}).$$

By *A18* and the *freshness rule*, *Statement 34* can be derived.

$$G| \equiv \#(< ID_{UAV}, R_{UAV}, T_{UAV2} >_{PK_{GCS}}, < \qquad \text{(Statement 34)}$$
$$H(SEK_{GU}, T_{GCS2}) >_{CHK_{Gu}}).$$

By (*Statement 33*), *A20*, and the *message meaning rule*, *Statement 35* can be derived.

$$G| \equiv U| \sim (< ID_{UAV}, R_{UAV}, T_{UAV2} >_{PK_{GCS}}, < \qquad \text{(Statement 35)}$$
$$H(SEK_{GU}, T_{GCS2}) >_{CHK_{Gu}}).$$

By (*Statement 34*), (*Statement 35*), and the *nonce verification rule*, *Statement 36* can be derived.

$$G| \equiv U| \equiv (< ID_{UAV}, R_{UAV}, T_{UAV2} >_{PK_{GCS}}, < \qquad \text{(Statement 36)}$$
$$H(SEK_{GU}, T_{GCS2}) >_{CHK_{Gu}}).$$

By (*Statement 36*) and the *belief rule*, *Statement 37* can be derived.

$$G| \equiv U| \equiv U \overset{SEK_{Gu}}{\leftrightarrow} G. \qquad \text{(Statement 37)}$$

By (*Statement 37*), *A22*, and the *jurisdiction rule*, *Statement 38* can be derived.

$$G| \equiv U \overset{SEK_{Gu}}{\leftrightarrow} G. \qquad \text{(Statement 38)}$$

By (*Statement 38*) and the *belief rule*, *Statement 39* can be derived.

$$G| \equiv U| \equiv ID_{UAV}. \qquad \text{(Statement 39)}$$

By (*Statement 39*), *A24*, and the *jurisdiction rule*, *Statement 40* can be derived.

$$G| \equiv ID_{UAV}. \qquad \text{(Statement 40)}$$

f The UAV U authenticates the ground control station G.

By *M6* and the *seeing rule*, *Statement 41* can be derived.

$$U \lhd (< ID_{GCS}, R_{GCS}, T_{GCS2} >_{PK_{UAV}}, < \qquad \text{(Statement 41)}$$
$$H(SEK_{GU}, T_{UAV2}) >_{CHK_{UG}}).$$

By *A17* and the *freshness rule*, *Statement 42* can be derived.

$$U| \equiv \#(< ID_{GCS}, R_{GCS}, T_{GCS2} >_{PK_{UAV}}, < \qquad \text{(Statement 42)}$$
$$H(SEK_{GU}, T_{UAV2}) >_{CHK_{UG}}).$$

By (*Statement 41*), *A19*, and the *message meaning rule*, *Statement 43* can be derived.

$$U| \equiv G| \sim (< ID_{GCS}, R_{GCS}, T_{GCS2} >_{PK_{UAV}}, < \qquad \text{(Statement 43)}$$
$$H(SEK_{GU}, T_{UAV2}) >_{CHK_{UG}}).$$

By (*Statement 42*), (*Statement 43*), and the *nonce verification rule*, *Statement 44* can be derived.

$$U| \equiv G| \equiv (< ID_{GCS}, R_{GCS}, T_{GCS2} >_{PK_{UAV}}, < \qquad \text{(Statement 44)}$$
$$H(SEK_{GU}, T_{UAV2}) >_{CHK_{UG}}).$$

By (*Statement 44*) and the *belief rule*, *Statement 45* can be derived.

$$U| \equiv G| \equiv U \overset{SEK_{Gu}}{\leftrightarrow} G. \qquad \text{(Statement 45)}$$

By (*Statement 45*), *A21*, and the *jurisdiction rule*, *Statement 46* can be derived.

$$U| \equiv U \overset{SEK_{Gu}}{\leftrightarrow} G. \qquad \text{(Statement 46)}$$

By (*Statement 46*) and the *belief rule*, *Statement 47* can be derived.

$$U| \equiv G| \equiv ID_{GCS}. \qquad \text{(Statement 47)}$$

By (*Statement 47*), *A23*, and the *jurisdiction rule*, *Statement 48* can be derived.

$$U| \equiv ID_{GCS}. \qquad \text{(Statement 48)}$$

By (*Statement 38*), (*Statement 40*), (*Statement 46*), and (*Statement 48*), it can be proved that the UAV U and the ground control station G authenticate each other in the proposed scheme. Moreover, it can also be proved that the proposed scheme can establish a session key between the UAV U and the ground control station G.

In the proposed scheme, the ground control station authenticates the UAV by

$$CHK_{GU} \overset{?}{=} H_3(SEK_{GU}, T_{GCS2}).$$

If it passes the verification, the ground control station authenticates the legality of the UAV. The UAV authenticates the ground control station by

$$CHK_{UG} \overset{?}{=} H_3(SEK_{GU}, T_{UAV2}).$$

If it passes the verification, the UAV authenticates the legality of the ground control station. The player, UAV, and ground control station authentication and communication phase of the proposed scheme thus guarantees mutual authentication between the UAV and the ground control station.

Scenario: A malicious attacker uses an illegal mobile reader to control an UAV.
Analysis: The attacker will not succeed because the illegal mobile reader has not been registered to the trusted authority center and thus cannot calculate the correct session key SEK_{UP}. Thus, the attack will fail when the legal UAV attempts to authenticate the illegal mobile device. In the proposed scheme, the attacker cannot achieve their purpose using an illegal mobile device. In the same scenario, the proposed scheme can also defend against a malicious attack using an illegal ground control station to send a fake message to a legal UAV, because the illegal ground control station has not been registered to the trusted authority center and thus cannot calculate the correct session key SEK_{GU}. Thus, the attack will fail when the legal UAV attempts to authenticate the illegal ground control station.

4.2. Integrity and Confidentiality

To ensure the integrity and confidentiality of the transaction data, this study uses elliptic curve cryptography and Diffie–Hellman key exchange algorithm to calculate the session key SEK_{UP}, SEK_{GP} and SEK_{GU}, and also to protect the integrity and confidentiality. The malicious attacker cannot use the signatures (K_{UP1}, K_{UP2}), (K_{PU1}, K_{PU2}), (K_{GP1}, K_{GP2}), (K_{PG1}, K_{PG2}), (K_{GU1}, K_{GU2}), and (K_{UG1}, K_{UG2}) to calculate the correct session key SEK_{UP}, SEK_{GP}, and SEK_{GU}.

Only a legal mobile device or UAV can calculate the correct session key SEK_{UP}. The legal UAV calculates the session key

$$SEK_{UP} = H_2(K_{UP1}, K_{UP2})$$

and the legal mobile device calculates the session key

$$\begin{aligned}
SEK_{UP} &= H_2(K_{PU1}, K_{PU2}). \\
K_{PU1} &= S_{PMD}T_{UAV} + aPK_{UAV} \\
&= S_{PMD}bP + aS_{UAV}P \\
&= bS_{PMD}P + S_{UAV}aP \\
&= bPK_{PMD} + S_{UAV}T_{PMD} = K_{UP1} \\
K_{PU2} &= aT_{UAV} = abP = baP = bT_{PMD} = K_{UP2}
\end{aligned}$$

Only a legal mobile device or ground control station can calculate the correct session key SEK_{GP}. The legal ground control station calculates the session key

$$SEK_{GP} = H_2(K_{GP1}, K_{GP2})$$

and the legal mobile device calculates the session key

$$\begin{aligned}
SEK_{UP} &= H_2(K_{PU1}, K_{PU2}). \\
K_{PG1} &= S_{PMD}T_{GCS} + cPK_{GCS} \\
&= S_{PMD}dP + cS_{GCS}P \\
&= dS_{PMD}P + S_{GCS}cP \\
&= dPK_{PMD} + S_{GCS}T_{PMD2} = K_{GP1} \\
K_{PG2} &= cT_{GCS} = cdP = dcP = dT_{PMD2} = K_{GP2}
\end{aligned}$$

Only a legal UAV or ground control station can compute the correct session key SEK_{GU}. The legal ground control station computes the session key

$$SEK_{GU} = H_2(K_{GU1}, K_{GU2})$$

and the legal UAV calculates the session key

$$SEK_{GU} = H_2(K_{UG1}, K_{UG2}).$$
$$K_{UG1} = S_{UAV}T_{GCS2} + ePK_{GCS}$$
$$= S_{UAV}fP + eS_{GCS}P$$
$$= fS_{UAV}P + S_{GCS}eP$$
$$= fPK_{UAV} + S_{GCS}T_{UAV2} = K_{GU1}$$
$$K_{UG2} = eT_{GCS2} = efP = feP = fT_{UAV2} = K_{GU2}$$

Only the correct session key will allow successful communication. Thus, attackers cannot decrypt or modify the transmitted message. Therefore, the proposed scheme achieves the integrity and confidentiality.

Scenario: A malicious attacker intercepts the transmitted message from the ground control station to the player and decrypts the message or sends a modified message to the player.
Analysis: The attacker will not succeed because the legal player will use

$$CHK_{PG} \overset{?}{=} H_3(SEK_{GP} \| T_{PMD2})$$

to check the integrity. The attacker cannot calculate the correct session key SEK_{GP}. Thus, the attack will fail when the legal player authenticates the received message. In the proposed scheme, the attacker cannot achieve his/her purpose by sending a modified message to the player, and he/she also cannot decrypt the intercepted message. For the same reason, the attack will fail when the legal ground control station uses

$$CHK_{GP} \overset{?}{=} H_3(SEK_{GP} \| T_{GCS})$$

to check the integrity. Therefore, attackers cannot achieve their purpose by sending a modified message to the ground control station or decrypt the intercepted message.

4.3. Identity Anonymity and Privacy

Another form of privacy attack involves attempting to obtain a player's real name or physical location by tracing his/her mobile device. If the mobile device sends the same message continuously, an attacker can trace its location. In the proposed scheme, the session key SEK_{UP} and SEK_{GP} is changed for every communication round in order to avoid location tracing. Besides, the pseudonym identity is used instead of real name in the proposed scheme. Thus, location privacy is protected and identity anonymity is achieved.

4.4. Availability and Prevention of DoS Attack

An attacker may impersonate a legal sender and then send the same message again to the intended receiver, trying to make the system unable to provide services properly. However, this attack will fail in the proposed scheme, as all messages between the sender and the receiver are protected with the session key SEK_{UP}, SEK_{GP}, and SEK_{GU}, and the attacker cannot calculate the correct session key. Because the transmitted messages are changed every round, the same message cannot be sent twice. Thus, the DoS attack is prevented and system availability is achieved.

4.5. Prevention of Spoofing Attack

In the proposed scheme, the GPS message is obtained by the UAV then transmitted to the ground control station or the player. The GPS message M_{GPS} is protected by the session key SEK_{UP} and SEK_{GU}. The attacker cannot compute the correct session key SEK_{UP} or SEK_{GU} and he/she cannot impersonate a legal UAV and send a fake message. Therefore, the spoofing attack is prevented.

Scenario: A malicious attacker pretends a legal UAV and sends a fake message to the legal ground control station.

Analysis: The attacker will not succeed because the illegal UAV has not been registered to the trusted authority center and thus cannot calculate the correct session key SEK_{GU}. Thus, the attack will fail when the legal ground control station attempts to authenticate the illegal UAV. In the proposed scheme, the attacker cannot achieve the purpose of pretending to be a legal UAV and sending a fake message. In the same scenario, the proposed scheme can also defend against a malicious attacker pretending to be a legal UAV and sending a fake message to the legal player, because the illegal UAV has not been registered to the trusted authority center and thus cannot calculate the correct session key SEK_{UP}. Thus, the attack will fail when the legal player attempts to authenticate the illegal UAV.

4.6. Non-Repudiation

In the proposed scheme, the digital signature is used to achieve non-repudiation between the parties in each phase. The sender uses his/her private key to sign the transmitted message, and the receiver uses the public key of the sender to verify the received message. Thus, the non-repudiation is achieved. Table 1 shows the non-repudiation of the proposed scheme.

Table 1. Non-repudiation of the proposed scheme.

Item / Phase	Proof	Issuer	Holder	Verification
Player and manufacturer authentication and communication phase	(C_{UAV}, Sig_{UAV})	M	P	$Sig_{UAV} = S_{SK_{UAV}}(M_{payment}, Cert_{UAV})$ $(M_{payment}, Cert_{UAV}) \overset{?}{=} V_{PK_{UAV}}(Sig_{UAV})$
Player and ground control station authentication and communication phase	(C_{GCS}, Sig_{GCS})	G	P	$Sig_{GCS} = S_{SK_{GCS}}(ID_{PMD}, M_{payment}, Cert_{UAV})$ $(ID_{PMD}, M_{payment}, Cert_{UAV}) \overset{?}{=} V_{PK_{GCS}}(Sig_{GCS})$
Player, UAV, and ground control station authentication and communication phase	(C_{PMD3}, Sig_{PMD3})	P	U	$Sig_{PMD3} = S_{SK_{PMD}}(M_{request}, Cert_{UAV})$ $(M_{request}, Cert_{UAV}) \overset{?}{=} V_{PK_{PMD}}(Sig_{PMD3})$
	(C_{GCS2}, Sig_{GCS2})	G	U	$Sig_{GCS2} = S_{SK_{GCS}}(ID_{PMD}, M_{confirm}, Cert_{UAV})$ $(ID_{PMD}, M_{confirm}, Cert_{UAV}) \overset{?}{=} V_{PK_{GCS}}(Sig_{GCS2})$
	(C_{UAV3}, Sig_{UAV3})	U	P	$Sig_{UAV3} = S_{SK_{UAV}}(ID_{PMD}, M_{confirm}, M_{GPS}, Cert_{UAV})$ $(ID_{PMD}, M_{confirm}, M_{GPS}, Cert_{UAV}) \overset{?}{=} V_{PK_{UAV}}(Sig_{UAV3})$
Ground control station and UAV authentication and communication phase	(C_{GCS3}, Sig_{GCS3})	G	U	$Sig_{GCS3} = S_{SK_{GCS}}(ID_{UAV}, M_{request})$ $(ID_{UAV}, M_{request}) \overset{?}{=} V_{PK_{GCS}}(Sig_{GCS3})$
	(C_{UAV4}, Sig_{UAV4})	U	G	$Sig_{UAV4} = S_{SK_{UAV}}(ID_{PMD}, M_{confirm}, M_{GPS}, Cert_{UAV})$ $(ID_{PMD}, M_{confirm}, M_{GPS}, Cert_{UAV}) \overset{?}{=} V_{PK_{UAV}}(Sig_{UAV4})$

4.7. Comparison of Security Issues

Table 2 shows a comparison of security issues of related works.

Table 2. Comparison of security issues.

	Yoon et al. [18]	Chen et al. [19]	Wazid et al. [20]	Tian et al. [21]	The Proposed Scheme
Mutual authentication	Unidirectional authentication	Yes	Yes	Unidirectional authentication	Yes
Integrity	N/A	Yes	No	Yes	Yes
Confidentiality	Yes	Yes	Yes	Yes	Yes
Identity anonymity	N/A	N/A	Yes	Yes	Yes
Availability	No	N/A	N/A	N/A	Yes
Privacy	N/A	N/A	Yes	Yes	Yes
Non-repudiation	No	Yes	No	Yes	Yes
DoS attack	Yes	N/A	Yes	N/A	Yes
Spoofing attack	N/A	N/A	Yes	N/A	Yes

4.8. Computation Cost

Table 3 shows the computation cost of the proposed scheme and Wazid et al.'s scheme [20].

T_P: Polynomial function operation
T_{Mul}:   Multiplication operation
T_H: Hash function operation
T_{Cmp}:   Comparison operation
T_{Enc}:   Symmetric encryption operation
T_{Sig}:   Signature operation
T_{Xor}:   Exclusive-or operation

Table 3. Computation cost of the proposed scheme and Wazid et al.'s scheme [21].

		Wazid et al. [20]	The Proposed Scheme
Manufacturer (UAV) registration phase	Manufacturer (UAV)	N/A	$2T_{Mul} + 1T_H + 1T_{Cmp}$
	Trusted authority center	$1T_P + 2T_H$	$2T_{Mul} + 1T_H$
Player (mobile device) registration phase	Player (mobile device)	$1T_P + 8T_H + 6T_{Xor}$	$2T_{Mul} + 1T_H + 1T_{Cmp}$
	Trusted authority center	$4T_H$	$2T_{Mul} + 1T_H$
Ground control station registration phase	Ground control station	N/A	$2T_{Mul} + 1T_H + 1T_{Cmp}$
	Trusted authority center	N/A	$2T_{Mul} + 1T_H$
Player and manufacturer authentication and communication phase	Player (mobile device)	N/A	$5T_{Mul} + 4T_H + 2T_{Cmp} + 2T_{Enc} + 1T_{Sig}$
	Manufacturer (UAV)	N/A	$5T_{Mul} + 4T_H + 1T_{Cmp} + 2T_{Enc} + 1T_{Sig}$
Player and ground control station authentication and communication phase	Player (mobile device)	N/A	$5T_{Mul} + 4T_H + 2T_{Cmp} + 2T_{Enc} + 1T_{Sig}$
	Ground control station	N/A	$5T_{Mul} + 4T_H + 1T_{Cmp} + 2T_{Enc} + 1T_{Sig}$
Player, UAV, and ground control station authentication and communication phase	Player (mobile device)	$1T_P + 16T_H + 3T_{Cmp} + 11T_{Xor}$	$1T_{Cmp} + 2T_{Enc} + 2T_{Sig}$
	Manufacturer (UAV)	$7T_H + 2T_{Cmp} + 4T_{Xor}$	$5T_{Mul} + 4T_H + 3T_{Cmp} + 4T_{Enc} + 3T_{Sig}$
	Ground control station	N/A	$5T_{Mul} + 4T_H + 1T_{Cmp} + 2T_{Enc} + 1T_{Sig}$
	Trusted authority center	$8T_H + 2T_{Cmp} + 5T_{Xor}$	N/A
Ground control station and UAV authentication and communication phase	Ground control station	N/A	$1T_{Cmp} + 2T_{Enc} + 2T_{Sig}$
	Manufacturer (UAV)	N/A	$1T_{Cmp} + 2T_{Enc} + 2T_{Sig}$

In Table 3, computation costs of the proposed scheme and Wazid et al.'s for the trusted authority center, manufacturer (UAV), player (mobile device), and ground control station in each phase are analyzed. For the highest computation cost in the player, UAV, and ground control station authentication and communication phase, a UAV needs five multiplication operations, four hash function operations, three comparison operations, four symmetric encryption operations, and three signature operations. A player needs one comparison operation, two symmetric encryption operations, and two signature operations. A ground control station needs five multiplication operations, four hash function operations, one comparison operation, two symmetric encryption operations, and one signature operation. The computation cost is acceptable in the proposed scheme.

4.9. Communication Cost

The communication cost of the proposed scheme and Wazid et al.'s scheme [20] is shown in Table 4.

Table 4. Communication cost of the proposed scheme and Wazid et al.'s scheme [21].

		Wazid et al. [20]	The Proposed Scheme
Manufacturer (UAV) registration phase	Message length	560 bits	2528 bits
	Round	1	2
	3.5G (14 Mbps)	0.040 ms	0.181 ms
	4G (100 Mbps)	0.006 ms	0.025 ms
Player (mobile device) registration phase	Message length	880 bits	2528 bits
	Round	2	2
	3.5G (14 Mbps)	0.063 ms	0.181 ms
	4G (100 Mbps)	0.009 ms	0.025 ms
Ground control station registration phase	Message length	N/A	2528 bits
	Round	N/A	2
	3.5G (14 Mbps)	N/A	0.181 ms
	4G (100 Mbps)	N/A	0.025 ms
Player and manufacturer authentication and communication phase	Message length	N/A	2816 bits
	Round	N/A	4
	3.5G (14 Mbps)	N/A	0.201 ms
	4G (100 Mbps)	N/A	0.028 ms
Player and ground control station authentication and communication phase	Message length	N/A	2816 bits
	Round	N/A	4
	3.5G (14 Mbps)	N/A	0.201 ms
	4G (100 Mbps)	N/A	0.028 ms
Player, UAV, and ground control station authentication and communication phase	Message length	1840 bits	5536 bits
	Round	3	6
	3.5G (14 Mbps)	0.131 ms	0.395 ms
	4G (100 Mbps)	0.018 ms	0.055 ms
Ground control station and UAV authentication and communication phase	Message length	N/A	2720 bits
	Round	N/A	2
	3.5G (14 Mbps)	N/A	0.194 ms
	4G (100 Mbps)	N/A	0.027 ms

The communication efficiency of the proposed scheme and Wazid et al.'s scheme during the transaction process of each phase was also analyzed. It was assumed that an elliptic curve modular operation required 160 bits, a hash operation required 160 bits, an AES operation required 256 bits, a signature operation required 1024 bits, and other messages, such as id, pid, and random number, required 80 bits. For example, the player, UAV and ground control station authentication and communication phase of the proposed scheme requires four elliptic curve modular messages, two hash messages, four AES messages, three signature operation messages, and six other messages. It thus requires $160 \times 4 + 160 \times 2 + 256 \times 4 + 1024 \times 3 + 80 \times 6 = 5536$ bits. In a 3.5G environment, the maximum transmission speed is 14 Mbps. This study also considered the player, UAV, and ground control station authentication and communication phase of the proposed scheme, which only takes 0.395 ms to transfer all messages. In a 4G environment, the maximum transmission speed is 100 Mbps and the transmission time is reduced to 0.055 ms.

Basically, Wazid et al.'s scheme provides a lightweight user authentication scheme in which a user in the IoD environment needs to access data. This appeals as it aims at providing a fast authorization mechanism. However, the integrity, non-reputation, and availability issues are excluded. However, compared to Wazid et al.'s scheme, the proposed scheme used the public key cryptography to design a UAV application field which was applied in a sensitive field such that the integrity, non-reputation and availability issues needed to be considered and should be ensured [20]. The proposed scheme is a different application field to Wazid et al.'s scheme. The players must pass necessary procedures to obtain the flight authority in a sensitive area. It needs more scenarios and overloads. As shown in Table 4, the communication cost sounds good. The proposed scheme provides a novel solution in the UAV application field.

Compared to the Wazid et al.'s scheme, the proposed scheme achieves the following advantages: firstly, the proposed scheme uses a signature mechanism, thus it can ensure data integrity and achieve non-repudiation and secondly, the proposed architecture involves the role of the ground control station to effectively grasp the UAVs' flying status in a sensitive area. The ground control station can also confirm whether the flying UAV is authorized. Although the proposed architecture has higher computing and communication costs than the Wazid et al.'s scheme, it also achieves higher security and availability.

5. Conclusions

At present, UAVs are mainly used for small package delivery and leisure entertainment. In the future, they will have thousands of uses that could even be widely extended to agricultural, land protection surveillance, emergency relief, military reconnaissance, space exploration, and other applications. UAVs will also create new jobs, while also addressing population ageing and manpower shortages. Advanced technology can bring a better and convenient living environment for mankind, but UAVs can also be maliciously used, and even endanger national security.

In this paper, a traceable and privacy protection protocol was designed to conduct the UAVs' application in sensitive control area. The proposed scheme creates a feasible and secure management platform in a sensitive area surveillance for UAVs' application. For sensitive military areas, players must obtain flight approval from a ground control station before they can control the UAV in these sensitive areas. The proposed scheme achieves mutual authentication, integrity and confidentiality, anonymity and privacy, non-repudiation, availability and protection against DoS attack, while also preventing spoofing attack. This study also analyzed the computation cost and the communication cost in the proposed scheme to prove the proposed scheme is practical in the real world.

Author Contributions: Conceptualization, Y.-Y.D. and C.-L.C.; methodology, Y.-Y.D. and C.-L.C.; validation, W.W., C.-H.C., Y.-J.C., and C.-M.W.; investigation, W.W. and C.-H.C.; data analysis, C.-H.C., Y.-J.C., and C.-M.W.; writing—original draft preparation, Y.-Y.D.; writing—review and editing, C.-L.C.; supervision, C.-L.C. and C.-H.C. All authors have read and agreed to the published version of the manuscript.

Funding: This work was supported in part by the National Natural Science Foundation of China under Grant 61906043, Grant 61877010, Grant 11501114, and Grant 11901100, in part by the Fujian Natural Science Funds under Grant 2019J01243, and in part by Fuzhou University under Grant 510730/XRC-18075, Grant 510809/GXRC-19037, Grant 510649/XRC-18049, and Grant 510650/XRC-18050.

Conflicts of Interest: The authors declare no conflict of interest.

References

1. Maza, I.I.; Caballero, F.F.; Capitán, J.; Martínez-de-Dios, J.R.; Ollero, A. Experimental results in multi-UAV coordination for disaster management and civil security applications. *J. Intell. Robot. Syst.* **2011**, *61*, 563–585. [CrossRef]

2. Meng, X.; Wang, W.; Leong, B. SkyStitch: A cooperative Multi-UAV-based real-time video surveillance system with stitching. In Proceedings of the 23rd Annual ACM Conference on Multimedia Conference, Brisbane, Australia, 26–30 October 2015; pp. 261–270.

3. Sun, Z.; Wang, P.; Vuran, M.C.; Al-Rodhaan, M.A.; Al-Dhelaan, A.M.; Akyildiz, I.F. BorderSense: Border patrol through advanced wireless sensor networks. *Ad Hoc Netw.* **2011**, *9*, 468–477. [CrossRef]

4. Vollgger, S.A.; Cruden, A.R. Mapping folds and fractures in basement and cover rocks using UAV photogrammetry, cape liptrap and cape Paterson, Victoria, Australia. *J. Struct. Geol.* **2016**, *85*, 168–187. [CrossRef]

5. Cho, J.; Lim, G.; Biobaku, T.; Kim, S.; Parsaei, H. Safety and security management with unmanned aerial vehicle (UAV) in oil and gas industry. *Proc. Manuf.* **2015**, *3*, 1343–1349. [CrossRef]

6. Zaouche, L.; Natalizio, E.; Bouabdallah, A. ETTAF: Efficient target tracking and filming with a flying ad hoc network. In Proceedings of the 1st ACM International Workshop on Experiences with the Design and Implementation of Smart Objects, Paris, France, 7 September 2015; pp. 49–54.

7. Danoy, G.; Brust, M.R.; Bouvry, P. Connectivity stability in autonomous multi-level UAV swarms for wide area monitoring. In Proceedings of the 5th ACM Symposium on Development and Analysis of Intelligent Vehicular Networks and Applications, Cancun, Mexico, 2–6 November 2015; pp. 1–8.

8. Ben-Asher, Y.; Feldman, S.; Gurfil, P.; Feldman, M. Distributed decision and control for cooperative UAVs using ad hoc communication. *IEEE Trans. Control Syst. Technol.* **2008**, *16*, 511–516. [CrossRef]

9. Nader, M.; Jameela, A.J.; Imad, J. Unmanned aerial vehicles applications in future smart cities. *Technol. Forecast. Soc. Chang.* **2018**, in press. [CrossRef]

10. Sedjelmaci, H.; Senouci, S.M.; Messous, M. How to Detect Cyber-Attacks in Unmanned Aerial Vehicles Network? In Proceedings of the 2016 IEEE Global Communications Conference (GLOBECOM), Washington, DC, USA, 4–8 December 2016; pp. 1–6. [CrossRef]

11. Chriki, A.; Touati, H.; Snoussi, H.; Kamoun, F. FANET: Communication, mobility models and security issues. *Comput. Netw.* **2019**, *163*, 106877. [CrossRef]

12. Strohmeier, M.; Lenders, V.; Martinovic, I. On the security of the automatic dependent surveillance-broadcast protocol. *IEEE Commun. Surv. Tutor.* **2015**, *17*, 1066–1087. [CrossRef]

13. Wesson, K.D.; Humphreys, T.E.; Evans, B.L. Can Cryptography Secure Next Generation Air Traffic Surveillance? Technical Report. Available online: https://radionavlab.ae.utexas.edu/images/stories/files/papers/adsb_for_submission.pdf (accessed on 21 December 2019).

14. Sedjelmaci, H.; Senouci, S.M.; Ansari, N. Intrusion detection and ejection framework against lethal attacks in UAV-aided networks: A Bayesian game-theoretic methodology. *IEEE Trans. Intell. Transp. Syst.* **2016**, *18*, 1143–1153. [CrossRef]

15. Sedjelmaci, H.; Senouci, S.M.; Ansari, N. A hierarchical detection and response system to enhance security against lethal cyber-attacks in UAV networks. *IEEE Trans. Syst. Man Cybern. Syst.* **2017**, *48*, 1594–1606. [CrossRef]

16. García-Magariño, I.; Lacuesta, R.; Rajarajan, M.; Lloret, J. Security in networks of unmanned aerial vehicles for surveillance with an agent-based approach inspired by the principles of blockchain. *Ad Hoc Netw.* **2019**, *86*, 72–82. [CrossRef]

17. Xiao, L.; Xie, C.; Min, M.; Zhuang, W. User-centric view of unmanned aerial vehicle transmission against smart attacks. *IEEE Trans. Veh. Technol.* **2017**, *67*, 3420–3430. [CrossRef]

18. Yoon, K.; Park, D.; Yim, Y.; Kim, K.; Yang, S.K.; Robinson, M. Security authentication system using encrypted channel on UAV network. In Proceedings of the 2017 First IEEE International Conference on Robotic Computing (IRC), Taichung, Taiwan, 10–12 April 2017; pp. 393–398.
19. Chen, L.; Qian, S.; Lim, M.; Wang, S. An enhanced direct anonymous attestation scheme with mutual authentication for network-connected uav communication systems. *China Commun.* **2018**, *15*, 61–76. [CrossRef]
20. Wazid, M.; Das, A.K.; Kumar, N.; Vasilakos, A.V.; Rodrigues, J.J.P.C. Design and analysis of secure lightweight remote user authentication and key agreement scheme in internet of drones deployment. *IEEE Internet Things J.* **2018**, *6*, 3572–3584. [CrossRef]
21. Tian, Y.; Yuan, J.; Song, H. Efficient privacy-preserving authentication framework for edge-assisted Internet of Drones. *J. Inf. Secur. Appl.* **2019**, *48*, 102354. [CrossRef]
22. Han, W.; Zhu, Z. An ID-based mutual authentication with key agreement protocol for multiserver environment on elliptic curve cryptosystem. *Int. J. Commun. Syst.* **2014**, *27*, 1173–1185. [CrossRef]
23. Sarath, G.; Jinwala, D.C.; Patel, S. A Survey on Elliptic Curve Digital Signature Algorithm and its Variants. *Comput. Sci. Inf. Technol. (CSIT)–CSCP* **2014**, 121–136. [CrossRef]
24. Chen, L.; Morrissey, P.; Smart, N.P. Pairings in Trusted Computing. *LNCS* **2008**, *5209*, 1–17.
25. He, D.; Chan, S.; Guizani, M. Communication security of unmanned aerial vehicles. *IEEE Wirel. Commun.* **2017**, *24*, 134–139. [CrossRef]
26. Burrows, M.; Abadi, M.; Needham, R. A logic of authentication. *ACM Trans. Comput. Syst.* **1990**, *8*, 18–36. [CrossRef]

Article

A New FANET Simulator for Managing Drone Networks and Providing Dynamic Connectivity

Mauro Tropea *, Peppino Fazio, Floriano De Rango and Nicola Cordeschi

DIMES Department, University of Calabria, via P. Bucci 39/c, 87036 Arcavacata di Rende, Cosenza, Italy
pfazio@dimes.unical.it (P.F.); derango@dimes.unical.it (F.D.R.); nicola.cordeschi@unical.it (N.C.)
* Correspondence: mtropea@dimes.unical.it (M.T.); Tel.: +39-0984-494786

Received: 14 February 2020; Accepted: 19 March 2020; Published: 25 March 2020

Abstract: In the last decade, the attention on unmanned aerial vehicles has rapidly grown, due to their ability to help in many human activities. Among their widespread benefits, one of the most important uses regards the possibility of distributing wireless connectivity to many users in a specific coverage area. In this study, we focus our attention on these new kinds of networks, called flying ad-hoc networks. As stated in the literature, they are suitable for all emergency situations where the traditional networking paradigm may have many issues or difficulties to be implemented. The use of a software simulator can give important help to the scientific community in the choice of the right UAV/drone parameters in many different situations. In particular, in this work, we focus our main attention on the new ways of area covering and human mobility behaviors with the introduction of a UAV/drone behavior model to take into account also drones energetic issues. A deep campaign of simulations was carried out to evaluate the goodness of the proposed simulator illustrating how it works.

Keywords: FANET; coverage model; human mobility model; UAVs/drones positioning; energy model

---

1. Introduction

With the huge development of drone management, in terms of regulation [1] and ad-hoc protocols for 3D environments [2], Unmanned Aerial Vehicles (UAVs) are becoming a new and efficient way of providing wireless connectivity to the users in a specific geographical area covered by UAVs/drones. Given the possibility to access hostile environments and desolate places, drones can be used in many emergency situations (with the related disaster events) where the availability of temporary, prompt, and efficient communication with the outside world can provide important help for saving lives or executing rescue operations. Let us think, for example, to severe weather events, earthquakes, sabotage, and other natural or man-made disasters which can destroy water lines, roadways, bridges, oil and gas pipelines, power plants, transmission lines, and other infrastructures. The communication network composed by a dynamic set of UAVs is well known in the literature as Flying Ad-hoc NETwork (FANET), a particular type of Mobile Ad-hoc NETwork (MANET), in which the mobile nodes are the UAVs/drones that can construct self-organizing networks [3] and more complicated 3D mobility models. The way a fleet of drones can provide a reliable coverage service has been an important research topic for several years [4]. FANETs aim to maintain a certain quality of transmission deploying also node movement predictions [5–7]. The cooperation between these devices is fundamental for an optimal coverage service, because it is based on many coordination techniques, able to create collaborating fleet of drones, paying attention to energy issues [8] and channel state [9,10] conditions.

This paper is the extension of the conference paper [11] and it has the aim of providing the design and the realization of a software simulator written in Java language, able to point the attention on a set of important aspects of a FANET network. This kind of network is composed of a team of UAVs/drones that, as in our case, are able to provide and guarantee connectivity to users that are under their coverage. Moreover, the implemented software has the possibility of importing real maps directly from Google Maps platform. The particularity of the simulator is the management of the users based on a human mobility model that previews the subdivision of the map in different Points of Interest (PoIs) in which different typologies of users can move. PoIs represent the places crossed by mobile users that move towards a particular destination. The software is equipped also with a Graphical User Interface (GUI) that allows setting parameters such as the number of drones for covering an area, the number of users in the area, the specific drones parameters such as height and coverage radius, network resource in terms of available bandwidth, traffic typology, and so on. In addition to the previous conference paper, where the authors integrated the simulator with a *human mobility model*, for modeling users movements in the covered area, and a *footprint model*, for modeling drones channel and calculate the correct drone height on the basis of the coverage radius, a new *drone behavior model* taking into account the limited autonomy of these devices that have the necessity of recharging their batteries for continuing in the specific task and a *a energy consumption model* for considering the decreasing of energy during the flight are also introduced. The whole simulator was based on the current and newer Italian/European legislation related to the new ENAC proposal, which will be effective from July 2020, and on a review of the current state of UAV Regulations [12]. The constraints used in the simulator have been extracted from the LIC15 draft [13,14] and the main ones are resumed in Table 1. In the simulator, a standard link state protocol with periodical updates was implemented, for considering availability of bandwidth resources in the area where moving the users. Numerical results confirm the goodness of the proposed simulation software.

Table 1. Table of drones.

Subcategory	Operative Environment	Drone Class	Weight	Required Skills
A1 (VLOS, h<120m)	Overflight of uninvolved people (no gatherings)	C0 C1	<250 g; <900 g; (v<19 m/s, E<80J)	Online training and online exam prepared by the authority or center recognized by the authority
A2 (VLOS, h<120m)	50 m horizontal from uninvolved people, or 5 m if a low speed mode limited to 3 m/s is installed	C2	<4 kg	Certificate of competence issued by the authority or a center recognized by the authority, after: Online training and online examination as for A1. Self-practical training. Additional online training and online exam prepared by the authority or center recognized by the authority with additional subjects
A3 (VLOS, h<120m)	No presence of uninvolved people, 150 m from residential, commercial, industrial, or recreational areas	C3 C4	<25 kg <25 kg	Online training and online exam as for A1

The remainder of this paper is organized as follows. Section 2 presents the most recent works on the considered research topics. Section 3 describes the FANETSim network simulator, with more emphasis to its main components. Section 4 describes the Java simulation environment and the implementation details. Numerical results are presented in Section 5. Section 6 summarizes the main conclusions and future works.

2. Related Work

The UAVs management introduces many issues in terms of coverage guarantees, protocol communications, self coordination, path planning, software modules implementation, privacy, and civilian use [1]. It is a quite new opportunity for realizing a new kind of temporary and dynamic infrastructure and, in the last years, it has been the subject of a lot of research activity [2]. In fact, the utilization fields of these devices are many and different, such as: connectivity, precision agriculture, disaster and recovery, and so on. In the literature exist many studies concerning the use of UAVs/drones in different scenarios showing their enormous potentiality in many applicative domains. Among different works, some propose new software simulators tailored on specific applications and then capable of analyzing in depth the UAVs/drones' behavior. In the following, some of literature works are presented showing the studies of the researchers concerning drones' simulators suitable for covering a particular region and giving connectivity to the users.

2.1. UAVs/drones Path Planning

The general problem of the optimum path has been investigated for many years, not only in the UAV environment. However, when considering a particular architecture, the best-path evaluation from a source point to a destination point should consider some particular constraints.

For example, Huang et al. [4] analyzed the problem of drones deployment with particular attention to the maximization of terrestrial user coverage and the minimization of drone communication costs. The authors proposed an optimization model to seek the optimal positions on the street graph for the drones to optimize the objective subject to that the drones and the Base Stations form a connected graph. By the analysis of real datasets, the authors showed that a local optimum can be evaluated.

Mardani et al. [15] addressed the problem of path planning for maximizing the quality of video streaming applications. The authors proposed two schemes, based on the A* search algorithm with Dubins-path [16], taking into account the limited energy availability, the wind effect, and the path post-smoothing, in order to avoid squared paths. Both algorithms optimize the path jointly in terms of distance and throughput experienced by the drone. Numerical results confirm the effectiveness of the proposed scheme.

Ghaddar et al. [17] considered the management of UAV flights for monitoring geographical zones while optimizing the paths for covering the considered area of interests and minimizing energy consumption. The proposed scheme also aims to achieve the minimum completion-time, minimum number of turns, to lower the energy consumption, and having a shortest mission path to cover the whole area. They considered one rotary wing drone, without obstacles and non-flying zones. The path is planned again by considering a grid-based geographical decomposition.

2.2. UAV/Drone Simulators

In this subsection, we present some of the more recent works in literature that deal with UAV/drone simulators showing different contexts of use. The simulators are very important given that most of the experiments using real prototypes or systems are not feasible due to the costs and risks involved.

De Rango et al. [18–20] proposed a software simulator in the field of agriculture domain able to define specific variables and parameters and to test mechanisms of UAV/drone team control and coordination. These articles suggest new techniques of UAV coordination and monitoring of the specific region in order to cope with the parasites. The proposed techniques show how the UAVs' behavior and performance vary when different constraints, for example coverage range for communication, consumed energy, and drones resources, are taken into account.

Marconato et al. [21] proposed a hybrid software-based simulator called Aerial Vehicle Network Simulator (AVENS). This simulator merges X-Plane and OMNeT++ integrated with LARISSA (Layered architecture model for interconnection of systems in UAS). They integrated several networking protocols. The two simulators offer different functionalities: X-Plane allows controlling the flight; OMNeT++ allows measuring network parameters such as throughput, packet loss, and so on. Information between the two environments are exchanged through XML files.

Zema et al. [22] showed a novel simulation suite for networked flying robots, as the result of the integration of two already validated solutions: FL-AIR (Framework libre AIR) simulator and NS-3. The authors proposed a software solution that exploits the characteristics of these software to obtain a new one, called CUSCUS (CommUnicationS-Control distribUted Simulator), able to manage both networked and distributed control systems.

Baidya et al. [23] proposed a flexible and scalable open source simulator called FlyNetSim. Their software is able to simulate swarm of UAVs and uses two open source tools, namely ArduPilot and NS3, creating individual data paths between the devices operating in the system using a publish and subscribe-based middleware. They illustrated the capabilities of the proposal through cases of different scenarios.

Kate et al. [24] dealt with particular type of UAVs called micro-aerial vehicle (MAV) that are an emerging class of mobile sensing system. Their simulator, called Simbeeotic, allows considering swarms of MAVs and permits to model their key features, such as sensing and communication. The authors demonstrated that their software provides the appropriate level of fidelity to evaluate prototype systems while maintaining the ability to test at scale.

Javaid et al. [25] introduced in their paper a simulation testbed, called UAVSim (Unmanned Aerial Vehicle Networks cyber security analysis). The software allows users to easily experiment by adjusting different parameters for the networks, hosts and attacks. Each UAV device works on well-defined mobility framework and radio propagation models, which resembles real-world scenarios. Based on the experiments performed in their software, the authors evaluates the impact of Jamming attacks against UAV networks and reported the results to demonstrate the needing and usefulness of the testbed.

Al-Mousa et al. [26] proposed a simulator called UTSim able to simulate different UAV key aspects such as physical specification, navigation, control, communication, sensing, and avoidance in environments with static and moving objects in urban air traffic. The simulator is easy to use and permits specifying a set of parameters: the properties of the environment, the number and types of unmanned aerial vehicles in the environment, and the algorithm to be used for path planning and collision avoidance. The simulator is able to produce log files with a lot of useful information to be used for evaluating its goodness.

2.3. UAV/Drone Coverage

In this subsection, the studies on the coverage issues are analyzed. The research on connectivity is very interesting. The problems around connectivity, especially concerning the possibility of providing a prompt help in those situations of emergency such as catastrophic or disasters events, are an object of study by the research community.

Al-Hourani et al. [27] gave an interesting study on a RF propagation model for low altitude platforms providing a valid statistical approach. They proposed a simple analysis based on parameters belonging to the urban scenario that can be applied in those situation where the terrestrial infrastructure are "out-of-order" because of natural disaster. Their statistical model is able to perform prediction of path loss between the altitude platform and the terminals in a typical urban scenarios.

Park et al. [28] provided the results about their study on the possibility of giving aerial Wi-Fi network thanks to the use of UAVs/drones called Net-Drone. In this network, each aerial device acts as an access point for the users and it is able to give connection based on their mobility, which is especially useful for disaster areas where terrestrial network infrastructure is no longer available. Their proposal takes into account also the handover mechanisms, necessary to perform an adequate wireless coverage.

In addition, Sae et al. [29] analyzed, using low altitude platform, the possibility of providing a temporary communication network, composed of a number of devices equipped on board with dual technology, able to provide Wi-Fi connectivity acting as access point and a mesh network for the inter-devices communications. The authors provided a set of simulation results to prove the goodness of their proposal.

Xie et al. [30] analyzed and studied the optimal deployment density of Drone Small Cells (DSCs) to achieve maximum coverage considering the inter-cell interference. They split the channel propagation conditions as depending on two different signals, namely a probabilistic Line-of-Sight (LoS) and a probabilistic Non-Line-of-Sight (NLoS) links, which increases the computational difficulties in performance analysis. Their mathematical approach takes into account both LoS and NLoS links in order to consider the inter-cell interference issues affecting the communications.

2.4. Main Paper Contribution

This paper is the extension of the conference paper [11] where a new Java simulator has been proposed, able to simulate the dynamics of human movements and drones coverage. The main contribution of this paper is the introduction in the simulator of a new model for simulating drones coordination and behavior in order to re-accomplish the specific task and manage the charging operations. Thus, the developed simulator considers four important models implemented in the FANET network:

1. There is the model of a footprint, which links the coverage area, and then the radius and height of these new devices.
2. The model of human behavior simulates real movements of users in the considered area.
3. The model of drones behavior in the sky considers their limited autonomy, and then the necessity of recharging their batteries for continuing in the specific task.
4. The model of energy consumption considers how the energy of a drone decreases during its flight.

In this work no collision issues are taken into account and thus no collision behavior is modeled. We considered energy consumption for the drone flight to model its limited autonomy and show the need to perform charging operations for guaranteeing service continuity.

The simulator is also able to collect some interesting statistics about the simulated entities, giving the possibility to analyze the critical aspects that could occur in real situations, in terms of bad service coverage or emergency situations.

Table 2 compares the drone simulators described in the related work Section 2.2 showing the implemented models for each simulator with respect to the models implemented in our simulator software.

Table 2. Comparison of UAVs software simulators on implemented models.

Simulator	Applications	Human Mobility Model	Coverage Model Model	Path Planning Model	Energy Model
AVENS [21]	R&D	-	-	x	-
CUSCUS [22]	R&D	-	-	x	-
FlyNetSim [23]	R&D	-	x	x	-
Simbeeotic [24]	R&D	-	x	-	-
UAVSim [25]	R&D	-	x	x	-
UTSim [26]	R&D	-	x	x	-
FANETSim	R&D	x	x	x	x

3. FANETSim: FANET Simulation System

The developed simulator is Java software able to consider a set of UAVs/drones that fly in the sky for providing connectivity to the users inside the considered map, as shown in Figure 1. They have to be able to communicate with each other in order to exchange protocol messages and they have to guarantee connectivity to the users that require resources for a specific application. The simulator is equipped with a GUI that gives the possibility of setting the opportune values of the considered input parameters and, moreover, it permits extracting from Google Maps the real map to be considered for the simulations. In this map, it is possible to analyze the human behavior on the basis of the considered human mobility model that gives some indications on how people move during the day.

For the simulation, two different types of applications were considered: video and audio streaming. The main actors of the simulator are represented by the two main nodes: *UAVs/drones* and *Users*. In the following, the models considered for the aerial devices and for the people are shown.

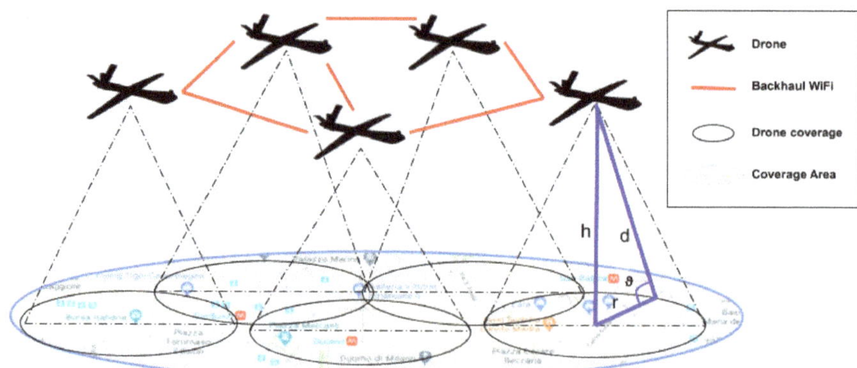

Figure 1. UAV/drone coverage footprint on the map.

As illustrated below, the FANET network topology is modeled by a grid graph structure, in which each node represents a drone and the links represent the connections between them (nodes regarding users are neglected in the forwarding network topology). The obtained adjacency matrix related to the graph is typically sparse, but it is highly dynamic given the mandatory drones mobility (from/to charging stations). The protocol messages exchanged by nodes (UAVs/drones and Users) allow the set up of the FANET network. Clearly, these nodes have to be inside a coverage that allows the communication between each devices. As mentioned above, the topology is modeled as a graph where an edge is represented by the couple of linked nodes. Each node (each vertex of the graph) has a list of neighbors that it stores in its own database.

3.1. Link State Protocol

The protocol used for the set up of the FANET network is a standard link state protocol that uses the typical messages preview by the standard. The link state protocol allows constructing the path between a couple of nodes defined source and destination. The protocol guarantees the possibility of topology adaptions, in case of topological changes when a new link is set up in the network or a link goes down. The simulator considers a real time delay in the communication, taking into account the propagation delay, the transmission delay, and the elaboration (processing) delay.

In the following, the messages sent for the creation of the FANET:

- **HELLO**: It is used for discovering the neighbor drones.
- **ACK**: It is used in reply to the HELLO message.
- **BYE**: It is used to disconnect a drone from the network.
- **LSA (Link State Advertisement)**: It carries the Link State Table and it is sent in flooding to all the neighbors.
- **LSU (Link State Update)**: It is used to update other routers with the information contained in the local router's database.
- **LU (Link Update)**: It is created to reduce traffic. Each drone sends it itself to account for its link update. The LU are grouped in a single LSU message and managed every 30 s. At every prefixed time, the drone generates a LSA packet with the updated information about its links.

The UAVs/drones can send the packets of the link state protocols containing the following main information: a number ID that identifies the node; a sequence number increased in each new packet;

and the Time To Live (TTL) of the packet. Moreover, the device uses a database for storing update information for each received LSA packet (see Figure 2).

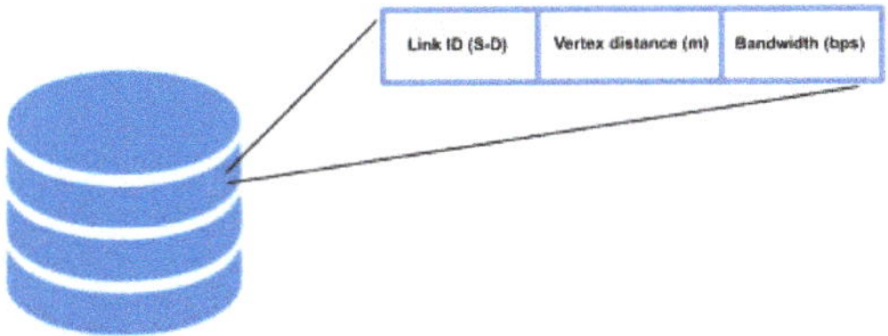

Figure 2. New record of LSA DB.

3.2. Human Mobility Model

The simulator introduces a human mobility model that helps to describe people behavior during the day giving indication on the main habits of the considered classes of users. This model is derived from a study conduced by the University of Milan [31]. They analyzed a smartphone call information dataset with over 69 million phone calls and 20 million text messages extracted in the city of Milan conduced for 67 continuous days. Moreover, in their work, the authors also studied the dataset of WiFi and GPS information based on the behavior of 178 people during four years (between 2007 and 2011). Thanks to this large volume of data, they extrapolated a human mobility model that considers three different categories of Points of Interest (PoIs) that represent the places visited by people during the day:

- **Mostly Visited PoI (MVP)**: locations most frequently visited by users (in this case are the homes and the workplaces);
- **Occasionally Visited PoI (OVP)**: locations of interest for the user, but visited just occasionally (in most cases, they correspond to favorite places or meeting points visited during the week); and
- **Exceptionally Visited PoI (EVP)**: locations that users rarely visit.

The behavior of users in the area has been considered and a classification of this people behavior has been introduced into the software simulator, considering the possible movement of the users in an urban scenario. Three different typologies of mobile users have been proposed:

- **Employee**: People who spend most of their time of a day at work. In addition, students can fit into this category.
- **Housewife**: People who spend most of their time, especially in the morning, at home.
- **Retired Person**: People who are free to move in the area without any constraints.

It is clear that each kind of users will spend its time during the day differently from other users categories, moving on the map with a different mobility model, as shown in Figure 3.

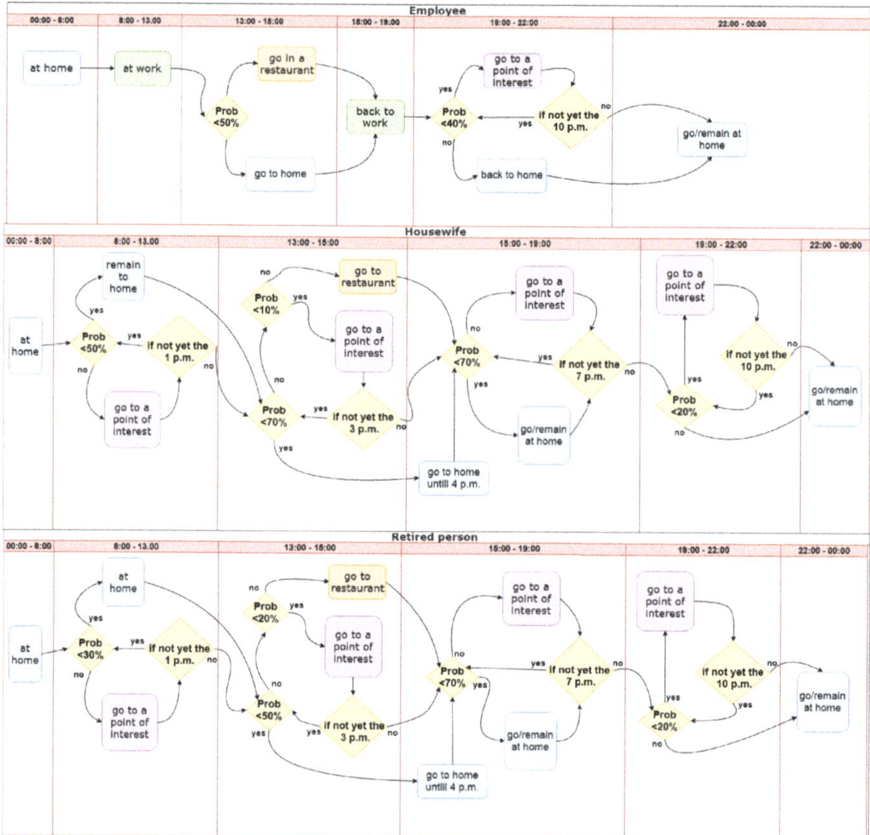

Figure 3. Diagram of human behavior for the different types of users.

The mobile user movements are influenced by their needs and social links. A user moves towards a series of places but he come back to home or to work that represent the most visited places. The probabilities associated with the most relevant places visited by users are computed on the basis of the visiting frequency. Equation (1) shows the probability that each mobile user m chooses of visiting the next $j - th$ Point of Interest, PoI_j, during a generic day, given that he has just visited the $i - th$ one:

$$P(PoI_{i,j}) = \frac{N_{vis}(PoI_{i,j})}{N_{visTOT_{out_i}}} \tag{1}$$

where $N_{vis}(PoI_{i,j})$ represents the number of times a mobile user has left PoI_i to go to PoI_j, and $N_{visTOT_{out_i}}$ is the total number of times a mobile user left PoI_i.

As regards the PoIs management with the mobility trace-files extracted from the considered map (Milan in our case), each mobile user chooses the next $j - th$ PoI to be visited probabilistically, on the basis of Equation (1). Our simulator gives the possibility to deploy a parametric number of PoIs as a square grid. Thus, without loss of generality, suppose we have a square geographical map with $M \times M$ size and a PoI grid equally spaced of $n \times n$ where each point represents the coordinates of a point of interest. The horizontal and vertical distance among PoIs will be M/n. Thus, if PoI_i has the coordinates (x_i, y_i), then its "pertinency" will occupy the area $B_i = [(x \pm Th_x), (y \pm Th_y)]$, where $Th_x = Th_y = \frac{M}{2n}$. To obtain the expression of Equation (1) a preliminary simulation round is necessary; the $j - th$ PoI is considered visited (and, hence, $N_{vis}(PoI_{i,j})$ is increased by 1 considering the starting

PoI_i) if users arrive inside B_i during the trips provided into the trace-files and remain inside the B_i for a certain temporal threshold indicated with parameter Th_t at least equal to 300 s [32]. In this way, Equation (1) can be defined for each defined PoI inside the map.

3.3. Footprint Coverage Model

In the considered urban scenario, a typical air to ground channel model has been considered composed mainly of two signals: Line of Sight (LoS) and a Non-Line of Sight (NLoS) [27]. These two typologies of signal can be considered separately and, then, they are studied considering different occurrence probabilities, which are functions of environment, density and height of buildings, and elevation angle. In this context, the impact of small scale fading has been neglected considering that fading probability is much smaller that the probability of receiving LoS and NLoS components [33]. In these conditions, for NLoS connections affected by shadowing and signals obstacles reflection, Path Loss (PL) is higher than LoS. For this reason, additional path loss values are assigned to LoS and NLoS links in the free space propagation loss model. In Table 3, the mathematical formulation symbols used in this subsection are shown.

Table 3. Footprint coverage model: used symbols.

Symbols	Description
ξ_{LoS}	average additional loss to the free space propagation loss in LoS
ξ_{NLoS}	average additional loss to the free space propagation loss in NLoS
γ_{th}	SNR threshold
α	constant values which depend on the environment (rural, urban, dense urban, etc.)
β	constant values which depend on the environment (rural, urban, dense urban, etc.)
f_c	carrier frequency
PL	Path Loss
PL_{LoS}	Path Loss for the LoS component of the signal
PL_{NLoS}	Path Loss for the NLoS component of the signal
h	height
r	coverage radius
d	distance between drone and receiver
θ	elevation angle
$Pr(LoS)$	probability of having LoS connections at an elevation angle of θ
$Pr(NLoS)$	probability of having NLoS connections at an elevation angle of θ
P_{TX}	transmission power
P_{RX}	received power
N	noise power
μ	h on r ratio

A typical coverage area of a drone is characterized by h (height of the drone), r (coverage radius), $d = \sqrt{r^2 + h^2}$ (distance between user on the coverage edge and drone in the sky), and $\theta = tan^{-1}(h/r)$ (angle (in radiant) between r and d), as shown in Figure 1.

From the above, a formula for the Path Loss (PL) for LoS/NLoS conditions as in [27] is given:

$$PL_{LoS/NLoS}(dB) = 20log(4\pi f_c d/c) + \xi_{LoS/NLoS} \tag{2}$$

where $PL_{LoS/NLoS}$ is the average (PL) for LoS/NLoS links, $\xi_{LoS/NLoS}$ is the average additional loss to the free space propagation loss which depends on the environment, c represents the speed of light, and f_c is the carrier frequency. The probability of having LoS connections at an elevation angle of θ is given by [30]:

$$P_r(LoS) = \frac{1}{1 + \alpha \cdot exp[-(\beta[(180/\pi)\theta - \alpha])]} \tag{3}$$

where α and β are constant values which depend on the environment (rural, urban, dense urban, etc.).

The NLoS probability is $P_r(NLoS) = 1 - P_r(LoS)$. Equation (3) indicates that the LoS probability in the connection between drone and user is a function of θ. In particular, it expresses that by increasing the elevation angle θ, the shadowing effect decreases and clear LoS path exists with high probability. Finally, the average PL is given. It shows its dependence from altitude (h) and coverage radius (r):

$$\bar{PL}(r,h) = P_r(LoS) \cdot PL_{LoS} + P_r(NLoS) \cdot PL_{NLoS}. \tag{4}$$

On the basis of the previous drone channel model it is possible to provide a formula that links the optimal altitude h to the maximum ground coverage area of radius r. Considering the drone transmission power (P_{TX}), it is possible to compute the received power (P_{RX}) as: $P_{RX}(dB) = P_{TX} - PL(r,h)$

Having P_{RX}, an user inside the coverage area of a drone at an height of h can receive the signal if its Signal to Noise Ratio (SNR) is greater than a set threshold (γ_{th}):

$$\gamma(r,h) = \frac{P_{RX}}{N} > \gamma_{th} \tag{5}$$

where N is the noise power. From Equation (5), it is possible to state that, to find the maximum achievable coverage radius, it should be: $\gamma(r,h) = \gamma_{th}$. Setting transmission power, the optimal drone height for having maximum coverage is given by [33]:

$$\frac{180(\xi_{NLoS} - \xi_{LoS})\beta Z}{\pi(Z+1)^2} - \frac{20\mu}{log(10)} = 0 \tag{6}$$

where $Z = \alpha \cdot exp(-\beta[(180/\pi)tan^{-1}(\mu) - \alpha])$ and $\mu = h/r$. By solving Equation (6), h_{opt} and r_{max} are found. The following parameter values have been considered: f_c = 2 GHz; ξ_{LoS} = 1 dB; ξ_{NLoS} = 20 dB; α = 9.6; β = 0.28; and γ_{th} = 10 dB.

3.4. Drone Characteristics and Energy Model

In the simulator, a typology of a drone with a flight mechanism called *"fixed-wing"*, which is in the category of very small drones (with a weight between 100 g and 2 kg), has been considered. In particular, the model selected for this typology of drones is the "Parrot Disco", which has the following specific characteristics: a weight of about 750 g, a range of about 2 km for remote control, a flight time of about 45 min, and a speed of about 80 km/h. Moreover, it is equipped with a three-cell LiPo battery of about 2700 mAh/25 A. The typical applications for this typology of drones are recreation and connectivity [34].

The total energy consumption of the UAV includes two components. The first one is the communication-related energy, which is due to the radiation, signal processing, as well as other circuitry. The other component is the propulsion energy, which is required for ensuring that the UAV could remain in the sky for supporting its mobility, if needed. The energy of an aircraft is characterized by two parameters, which are specific energy distribution rate, driven by elevator, and total specific energy rate, driven by throttle [35]. Note that, in practice, the communication-related energy is usually much smaller than the UAV's propulsion energy consumption; thus, as shown in several studies, it is ignored. For the flight level with fixed altitude, the UAV's energy consumption can be considered only depending on the speed $v(t)$ and acceleration $a(t)$. For steady straight-and-level flight (SLF) with constant speed V, we have $||v(t)|| = V$ and $a(t) = 0, \forall t$. Thus, we have:

$$\bar{E}_{SLF}(V) = T \left(c_1 V^3 + \frac{c_2}{V} \right) \tag{7}$$

Equation (7) is the consumed energy of the flight device for a finite time horizon T, where c_1 and c_2 are two parameters that take into account: weight, wing area, air density, etc. For simplicity, in this work, the payload is considered into the overall weight of the aircraft and the UAV energy storage weight reduction over time (for consumed fuel or battery) is ignored. The energy consumption presented in Equation (7) is a function of V, and it consists of two terms: one is proportional to the cubed speed V; it is known as the *parasitic power* for overcoming the parasitic drag due to the aircraft's skin friction, form drag, etc. The other term, as shown in Equation (7), is inversely proportional to V; it is known as the *induced power* for overcoming the lift-induced drag, i.e., the resulting drag force due to wings redirecting air to generate the lift for compensating the aircraft's weight. Moreover, the formula shows the dependence of time T as the time.

3.5. UAV/Drone Behavior Model

In this section, the behavior of flying UAVs is described, as implemented in the proposed Java simulator. We based our implementation on a simplified version of the idea in [15], considering the energy-aware approach for generating optimal trajectories, neglecting wind conditions and considering a dynamic coverage range, in order to adapt the data transmission to clients needs, in terms of the number of service requests. In this paper, no collision avoidance issues are taken into account. We have considered that drones cannot be in the same airspace and no collision model has been implemented. These aspects will be considered in our future works. In Table 4, the symbols for the mathematical formulation used in this subsection are shown.

Table 4. Drone behavior model: used symbols.

Symbols	Description
e_j	edge $\in E$
E_{lm}	Energy consumed traveling between e_{jl} and e_{jm}
E_{res}	Residual Energy
E_{PNR}	Point of No Return (PNR) Energy
P_{sd}	Loopless path between a starting waypoint and a destination waypoint
PW_d	Consuming power
cs_j	$j - th$ charging station
t_{lm}	traveling time on (e_{j_l}, e_{j_m})

The authors of [15] based their path planning algorithm on the A* search algorithm and Dubins paths [16], with a computational complexity lower than the one of Dijkstra's algorithm, thanks to the use of a heuristic function and a smaller search space. It must be underlined that physical trajectories are computed off-line by internal controllers, while routing data distribution is based on the link state protocol, as described above. Internal trajectory algorithms are also responsible for compensating external disturbances of regular and stable movements and follow the indications evaluated by the path planning algorithm. Drones join the data network by exchanging the *HELLO* packets with their neighbors. Coverage range is dynamically adapted to cover bigger/smaller regions on the basis of the requested bandwidth and the number of users (this is the main relation with the human mobility model and PoIs). The coverage range is never lower than $h*$ (the distance from the drone and the terrestrial surface, in order to guarantee the air–terrestrial connectivity). Figure 4 represents a typical coverage scenario with only four drones.

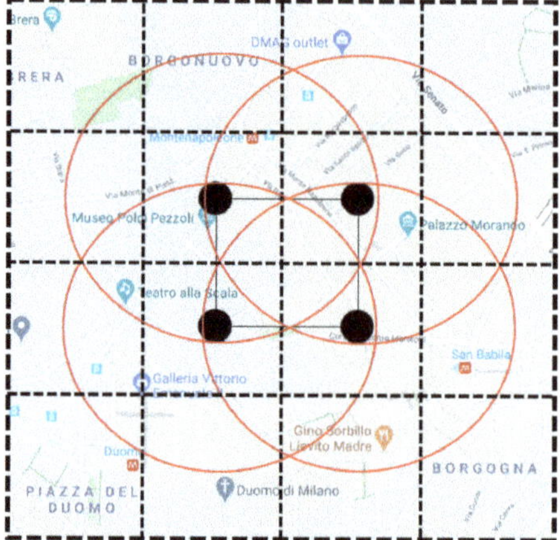

Figure 4. An example of geographical coverage: black points are the drones, thin circles represent their coverage area, and the link represent the connection for data/overhead/LSP transmission.

As illustrated in Figure 4, the flight area is modeled according to an undirected grid graph [17] $GG =< V, E >$, where $V = v_1, ..., v_n$ is the set of nodes (drones) associated to the n tiles and $E = e_1, ..., e_k$ is the set of edges. The number of drones n is equal to the number of tiles, while the number of edges k is variable and depending on the reciprocal coverage relationships. In fact, each drone is located at the center of each tile and its coverage range is directly proportional to the number of service requests of its tile. Differently from Mardani et al. [15], nodes corresponding to adjacent tiles are not surely connected by an edge: radiuses are adjusted on the basis of the bandwidth needs of the covered tile.

Each $v_i \in V$ is associated to a geographical position x_i, y_i and to a number of service requests s_i, while an edge $e_j \in E$ is associated to the radio connection (e_{j_l}, e_{j_m}) between nodes v_l and v_m. We do not consider obstacles in the flight area. Drones can fly from a tile to another one at constant speed v_d, while consuming a power PW_d. The initial energy available in each drone is E_s, while the energy consumption for traveling on (e_{j_l}, e_{j_m}) is E_{lm}. After the battery charging completion (in the simulator, different numbers of charging stations have been considered and they are placed on the map accordingly description in Section 4), each drone v_i starts its trip from the first tile near the charging station t_{sw} (starting waypoint) and it needs to arrive to the center of its associated tile t_{dw} (destination waypoint), by following a set of consecutive flying segments from t_{sw} to t_{dw}, forming a loopless path P_{sd}. If $PATH_{sd}$ is the set of all the possible loopless paths, the implemented drone mobility scheme, finds the optimal path $\in PATH_{sd}$ for minimizing energy consumption and respecting the following relationship:

$$\sum_{(e_{j_l}, e_{j_m}) \in P_{sd}} E_{lm} \leq E_{res} \tag{8}$$

with E_{res} residual energy of the drone (at the beginning $E_{res} = E_s$), while the energy consumption is $E_{lm} = PW_d \cdot t_{lm}$, with t_{lm} equal to the traveling time on (e_{j_l}, e_{j_m}). For simplicity, we did not considered the smoothing feature of Mardani et al. [15], thus drones move along the edges of the graph in a squared way.

The algorithm implemented in Java for the proposed simulator is composed of three main parts:

- *Initialization*: At the beginning, charging stations are placed as explained above and it is assumed that drones are fully charged, and then sent to the center of their own tile. When the drone arrives, the Point of No Return (PNR) energy E_{PNR} is evaluated for each drone, in order to know the amount of energy needed for having the possibility to go back to the charging station. Each drone then adjusts its coverage radius on the basis of the number of the number of service requests and starts the flooding of *HELLO* messages and the LSP. Once the grid graph (GG) is created, its adjacency matrix can be set as M.

- *Loop*: Each drone v_i has to continuously check its own residual energy level. When the energy drops down the E_{PNR} threshold, the drone in tile t_i needs to return to its charging station cs_j, thus the function *PLAN* is called to evaluate the best "go-back" path with the lowest energy consumption. If the drone v_i is in cs_j and its energy level has reached the maximum E_s, then the optimal path needs to be planned again, through the *PLAN* function.

- *PLAN*: This function receives as inputs GG, source and destination nodes, and it returns the evaluated path in the variable p. Its structure is very similar to the typical minimum search *Dijkstra's* algorithm, but the *extract-min-energy* function takes into account the relationship of Equation (8). In fact, the energy needed by a drone to traverse an edge e_{lm} is given by $E_{lm} = PW_d \cdot t_{lm}$ and it is taken into account by this function as the metric for each edge of the graph.

4. UAV/Drone Simulation Environment

The geographical grid model has been based on the center of *Milan*, with its center GPS coordinates equal to 45.468197, 9.193153. The map was extracted from Google Maps platform considering real coordinates. Several PoIs (shops, restaurants, hotels, and churches) have been taken into account, as well as people's homes (a grid of $12 \times 12 = 144$ PoIs was chosen to cover most of the map).

Pedestrian mobility was then generated as follows: each mobile user will start its activity at midnight of a generic working day, then it will stay at home until 8:00 am. After this time, all users will move during the day according to their own behavior, as shown in Figure 3. There are three types of moving users (employees, housewives and retired persons), by defining their percentage referring to the total number of moving nodes. To obtain deterministic human movements, a random seed is extracted and it is used to choose the next location probabilistically. Then, each mobile human node will reach it by following a real path (the best one existing into the extracted geographical map). The best path is obtained by searching, each time, for a next PoI of crossing. A grid on the map with a radius of about 80 m between two points has been considered.

Figure 5 explains better this motivation by an example: a user that goes from point S towards point D can choose one of Points 1–3. Among these, he chooses Point 2 because it satisfies the conditions of the algorithm: it is the point reachable from S closer to D.

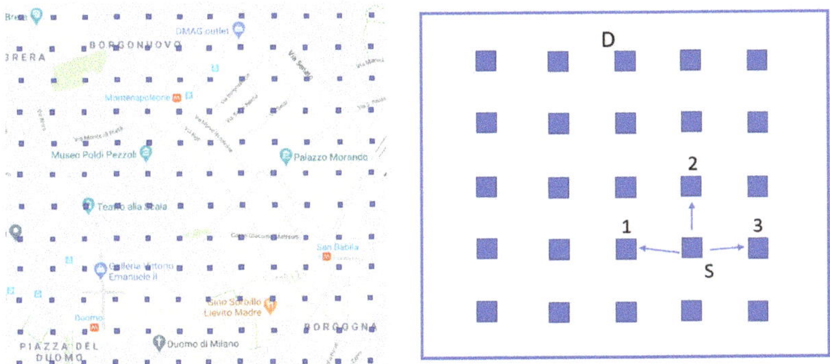

Figure 5. Example of next PoI chosen by user to reach point D from S (Point 2 is the closer to D).

Each human node reaches its destination by moving at a speed of 3 km/h. If, after a moving step, it still does not reach the destination, then the next PoI is recursively re-evaluated, and another moving step is generated. On the contrary, if the destination has been reached, then two other events related to the pause of people's mobility are initiated: a person who is not working (or is not at home) can stay in a particular PoI for a time belonging to the interval [10, 60] minutes. All activities end at 10:00 pm, and after this time, no more activities are provided (all the humans will return at home).

As regards flying devices, in the software simulator we considered the charging stations placement as illustrated in Figure 6: four stations are placed at the four corners of the simulated map.

When a flying drone needs to go to the assigned charging station to charge its battery, because it has reached the lower bound of residual energy, a backup drone is contacted before through a recruiting message (we assume that each flying drone knows exactly the address of its backup drone). Thus, the coverage hole is filled with the presence of the designated backup drone. We based our approach on the one specified in [20]. In our simulator, we have considered an approach based on drones recruiting. That is, to cope with the problem of the coverage hole left by a drone that needs to recharge its battery, we have considered, as some works in the literature suggest [36], the use of backup drones that lie near the charging station. This mechanism permits of replacing the drone that has to recharge its battery with a spare drone that is in an idle status, waiting for a recruiting message by a drone in the sky. The used policy for the recruitment is based on [20].

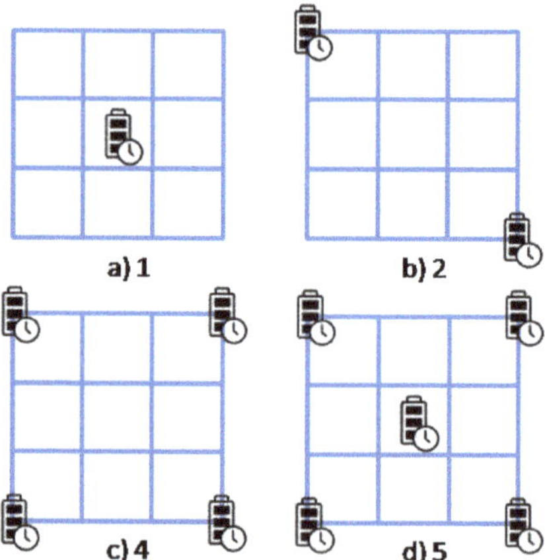

Figure 6. Charging stations placement with a typical UAV coverage.

The implemented simulator is able to simulate flying devices able to guarantee connectivity among peopel who can communicate using multimedia applications. In particular, two types of traffic have been considered: video and audio streaming.

The number of involved users is 50, 75, 100, 125, and 150. Each user is able to make a call each time it is in a PoI. Regarding the proposed human behavior model, 70% has been chosen for employees. Table 5 shows some of the considered parameters used in the simulative campaigns. For all simulations, a network topology consisting of $d = 9$ drones, with a distance of 252 m from each other, and disposed in a grid $d_x \times d_y = 3 \times 3$, has been used (this grid map permits of covering most of the considered area).

Table 5. Simulation parameters.

Parameter	Value
Simulation time	300 min
Area size	$M \times M = 1\,\mathrm{km} \times 1\,\mathrm{km} = 1\,\mathrm{km}^2$
Drones grid	$d_x \times d_y = 3 \times 3$
Drones resource	10 Mbps
Drones coverage radius	175 m
Drones distance	252 m
Drones optimal height	120 m
Drones buffer size	50 packets
Drones communication range	300 m
Recharging station number	1, 2, 4, 5
Backup Drones number	1, 2, 4, 5
PoIs grid	$n \times n$ with $n = 12$
PoIs distance	M/n meters $= 1000/12 \approx 80$ m
PoIs number	$n * n = 12 * 12 = 144$ (108 h 36 shops, restaurant, entertainment)
Th_x	$M/(2n) = 1000/24 \approx 40$ m
Th_y	$M/(2n) = 1000/24 \approx 40$ m
Th_t	300 s
Audio&Video Calls %	30% Audio & 70% Video
Users speed	3 km/h
Users number	50, 75, 100, 125, 150
Users' distribution	70% employees 20% housewives 10% retired persons

For all the simulations, a network topology consisting of nine drones, with a distance of 252 m each other and disposed in rows of three, has been considered (in this way the most of the area of interest is covered).

5. Performance Evaluation

Different simulation campaigns were performed, changing the number of users able to move in the considered geographical map. The following output parameters were collected by simulations:

- audio and video streams data delay;
- sent, received and lost packets;
- accepted and refused requests; and
- occupied bandwidth percentage.

These output parameters allowed evaluating the goodness of the proposed software simulator varying some input parameters.

5.1. UAVs/drones Simulation Tests without Energy Issues

In this subsection, we assume that each drone has unlimited energy availability, thus neither charging operations nor energy aware path planning are needed. Each flying device stays at the center of its grid element giving the needed coverage, in terms of bandwidth, while communicating with its neighbors via link state protocol. In this scenario, each UAVs/drones has no energy awareness and, then, only the network aspects are discussed with the considered graphics.

Figure 7a shows the curve trend of the sent and received packets in the scenario. It is possible to see how sent and received packets increase with the number of users on the map. This behavior of the simulator fits with the expected simulations results. This is the direct consequence of the available bandwidth on-board each device and the considered mobile users percentages among employees, housewives, and retired persons. Figure 7b shows the histogram of accepted requests, calls, and refused requests varying the number of user between 50 and 150. The trend of these results is in line with the expectation; in fact, the figure shows how the number of bandwidth requests increases with the number of users and how, therefore, calls and refused requests numbers increase. Figure 8 shows the trend of call admission behavior of each drone in the considered map. It shows the percentage of occupied bandwidth for each drone on the map and allows of making some consideration on the results. It is possible to view how Drones 3, 6, and 8 are the most loaded ones. This last ones present the greatest number of occupied bandwidth. In addition, the figure shows how the bandwidth percentage increases for higher number of users in the system and in the covered points of interest. In addition, it can be seen how the percentage of occupied bandwidth is under 26%, thus it results in a non-congested network, without any bottleneck in terms of throughput. Finally, the delay for audio and video streams is reported in the Figure 9. It shows the minimum, average, and maximum delay for both multimedia traffic, video and audio applications. In addition, for these simulation results, the obtained trend satisfies the expectations, respecting the typical values of these traffic typologies. Delay slightly increases for a higher number of mobile users, due to the higher protocol overhead.

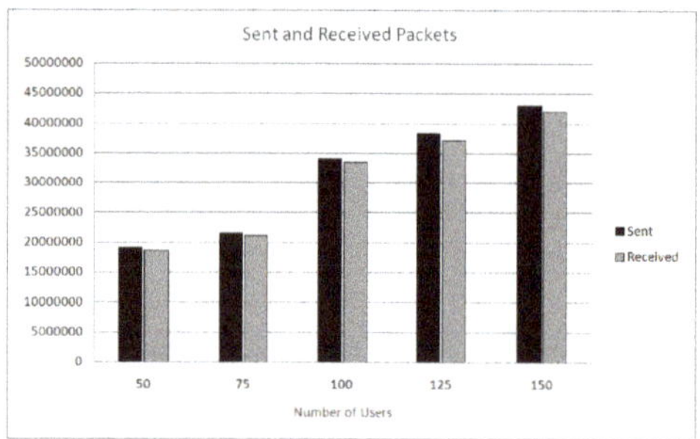

(a) Sent and received packets varying users number.

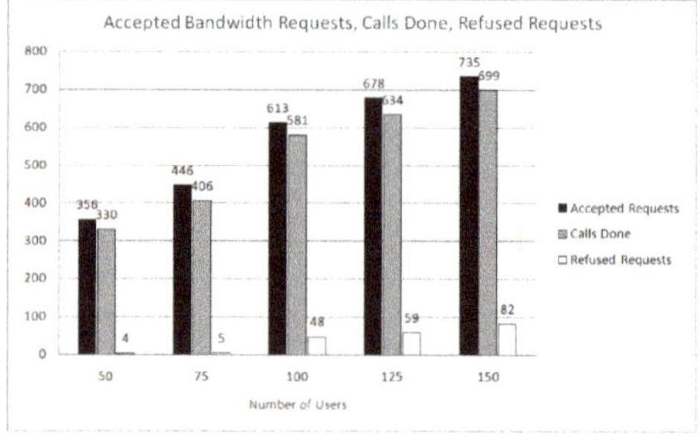

(b) Accepted requests, calls, refused requests varying users number.

Figure 7. Curves on packets and requests number.

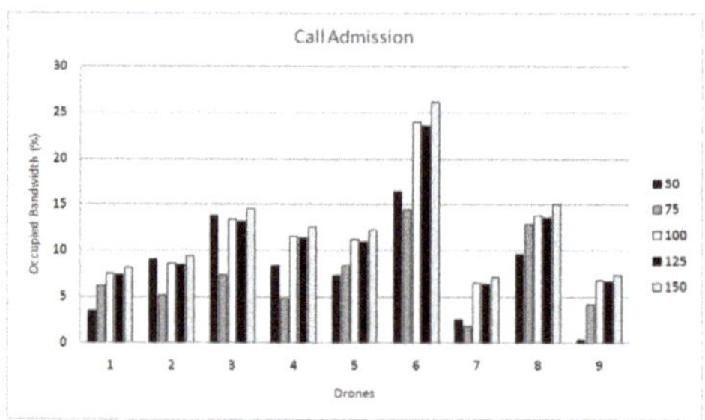

Figure 8. Drones call admission number varying users number.

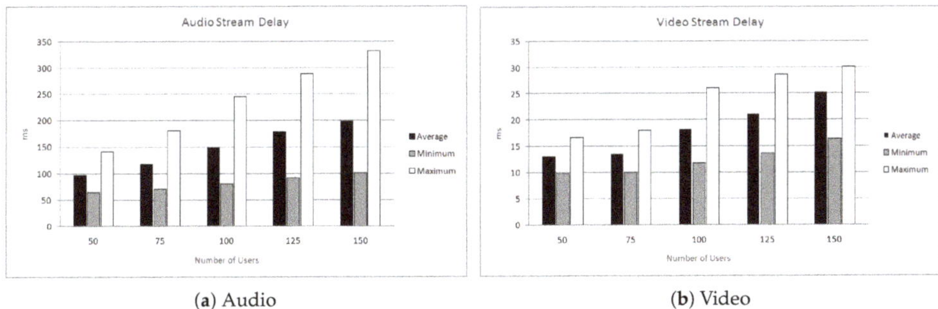

(a) Audio (b) Video

Figure 9. Audio and video stream delay (average, minimum, and maximum) varying users number.

5.2. UAVs/drones Simulation Tests with Energy Issues

Differently from the previous subsection, in this case, the residual energy is always taken into account. Thus, the energy aware path planning is mandatory to guarantee a proper coverage. In general, simulation results are worse than the previous scenario, because of the periodical need for charging operations: energy and time are wasted for flying from charging station to the grid elements and vice versa. In this scenario, firstly, we show some graphics considering four recharging stations without considering backup drones, and then we show a comparison considering 1, 2, 4, and 5 recharging stations with and without taking into account a backup drone for each station.

As for the previous case, also taking into account energy issues, we show the graphics for the sent and received packets in the overall considered system. Figure 10a shows the trend of the curves of sent and received packets highlighting the increasing behavior with the number of users on the map. This behavior of the simulator fits with the expected simulations results. Figure 10b shows that the number of accepted requests, calls, and refused requests are worse in this scenario. This type of behavior is due to the drones energy consumption. The drone that needs to recharge its battery recruits a new drone to permit the continuous coverage of its geographical area, guaranteeing the connectivity to the users connected; however, the drone exchange s takes some time, during which no connectivity can be provided. For this reason, a higher number of refused calls and a lower number of concluded calls are obtained. Figure 11 shows the average number of needed charges for each drone by varying the number of charging stations from one to five, as illustrated in Figure 6. As can be seen, the number of completed charging operations increases with the number of charging stations: this is due to the higher availability of charging chances. We assume that, if two or more drones compete for a charging place, they are served with a FIFO queue: waiting drones are placed near the charging stations (on the ground) and they are set in a stand-by mode (with a negligible energy consumption). Clearly, performance is degraded, in terms of not only energy consumption, but also served users, as can be seen in Figure 10b. In addition, for audio and video streaming delay, this scenario shows worse performance, as can be seen in Figure 12, because of the time needed for drones exchange and charging operations.

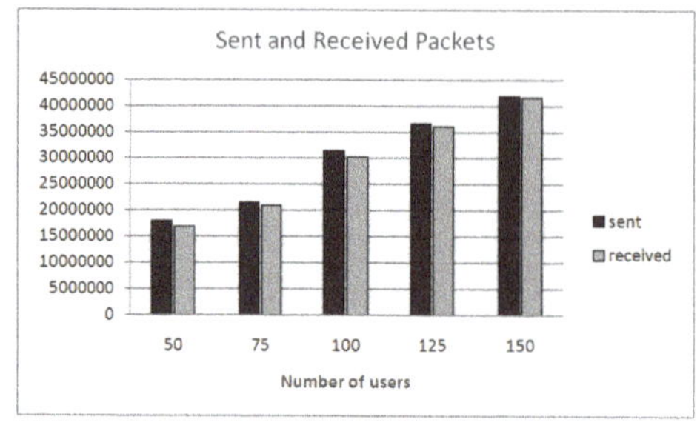

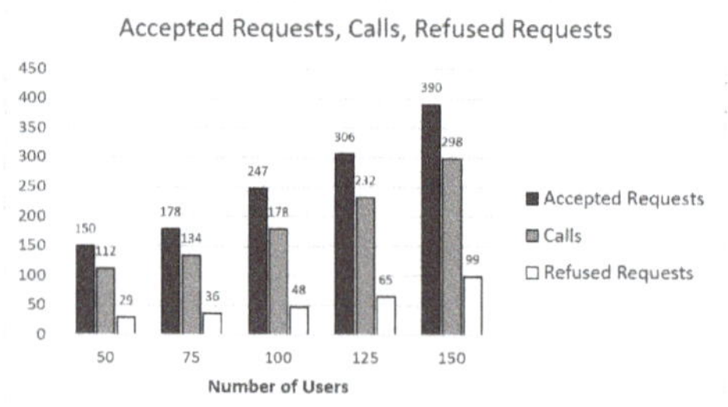

(**a**) Sent and received packets varying users number.

(**b**) Accepted requests, calls, refused requests varying users number.

Figure 10. Curves on packets and requests number.

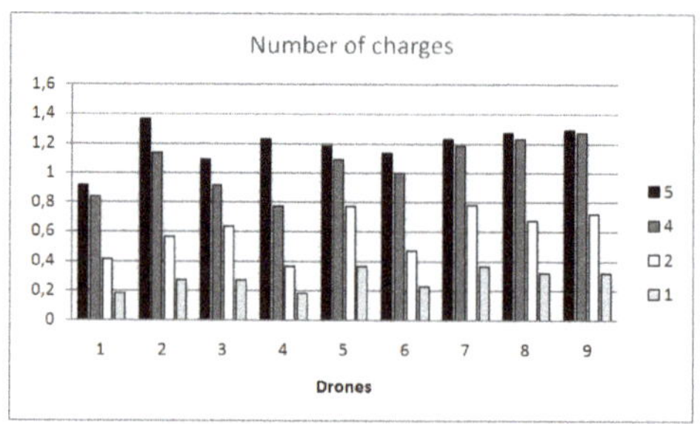

Figure 11. Number of charges varying charging stations number.

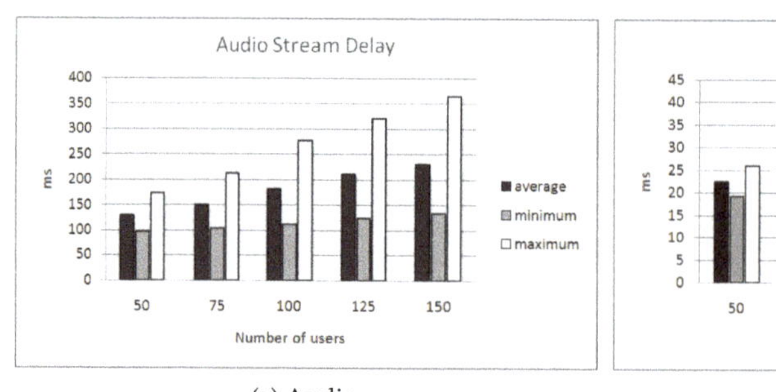

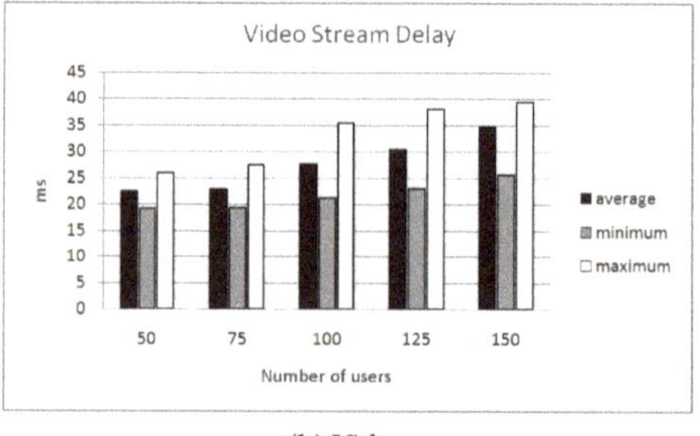

(**a**) Audio

(**b**) Video

Figure 12. Audio and video stream delay (average, minimum, and maximum) varying user number.

To show the effects of recharging stations in our simulator, we show some graphics about simulation results considering the use of 1–5 recharging stations on the map. Successively, we considered also the use of backup drones that can be recruited when a drone needs to perform recharging operations. The graphics in Figure 13 show how the introduction of recharging station to allow drones to recharge their battery results in worse system performance concerning the requests evaded by drones with respect to the case of considering no energy issues. The other series of graphics in Figure 14 show the performance in terms of accepted and refused requests when recharge stations and backup drones are implemented in the system together. The presence of spare drones allows replacing the device that has to recharge its battery. The drone that notices its battery level is low sends a recruiting message towards a drone placed near its recharging station. This drone reaches the location of the drone sending the request so it can provide transparently coverage to users.

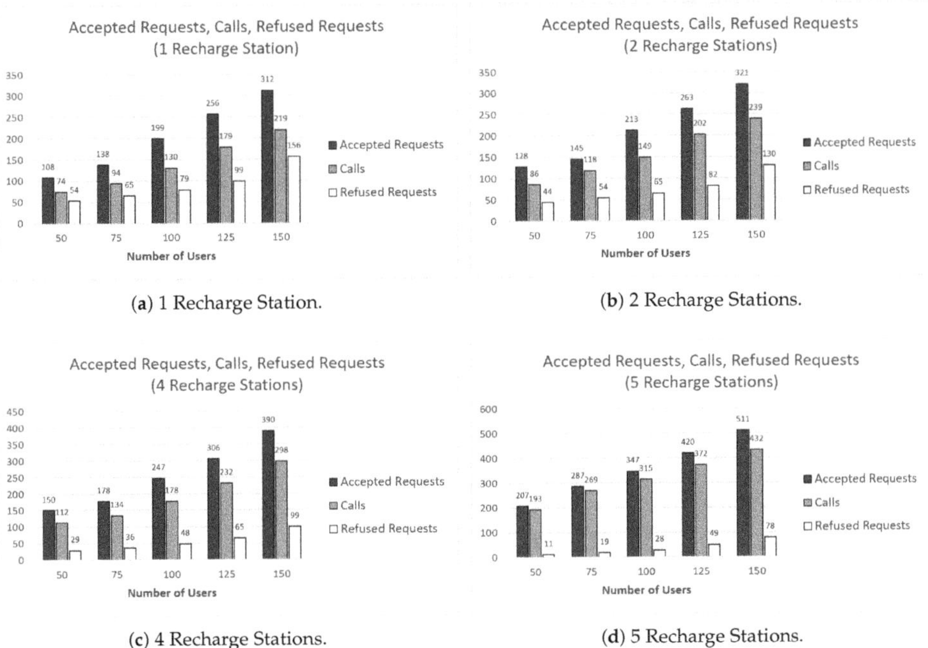

(**a**) 1 Recharge Station.

(**b**) 2 Recharge Stations.

(**c**) 4 Recharge Stations.

(**d**) 5 Recharge Stations.

Figure 13. Accepted requests, calls, and refused requests varying users and the recharge stations' number.

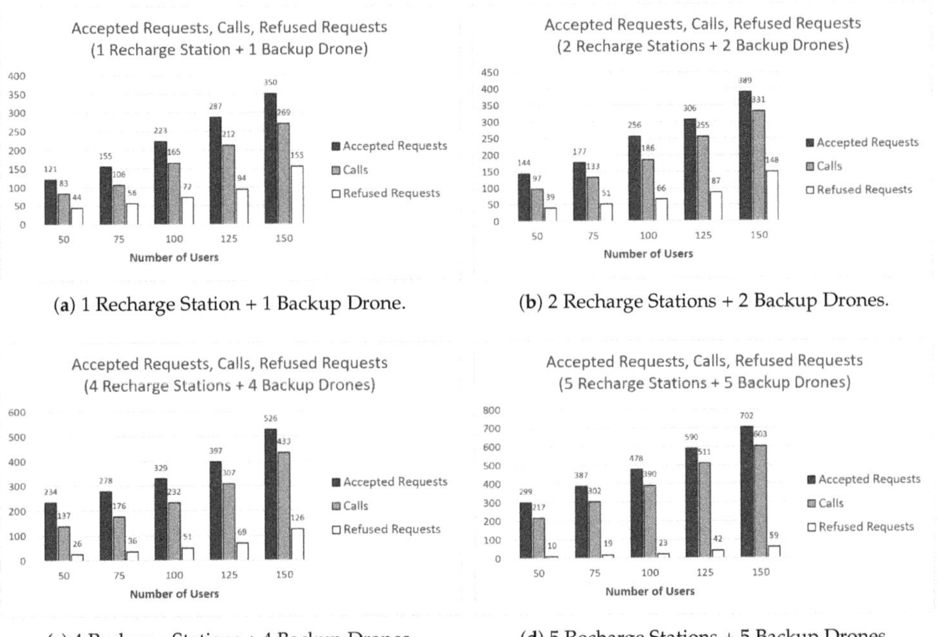

(**a**) 1 Recharge Station + 1 Backup Drone.

(**b**) 2 Recharge Stations + 2 Backup Drones.

(**c**) 4 Recharge Stations + 4 Backup Drones.

(**d**) 5 Recharge Stations + 5 Backup Drones.

Figure 14. Accepted requests, calls, and refused requests varying users and the recharge stations' number considering backup drones.

6. Conclusions

In this paper, a novel UAVs Java simulator is proposed. The main features of the simulator consist of the possibility of analyzing different aspects of the so-called FANETs. Flying devices (commonly defined UAVs or drones) are able to create temporary communications from air to ground, offering new connectivity, last-mile bandwidth distribution, guaranteeing efficient communications. The proposed Java software allows performing different simulation campaigns varying different input parameters. The simulator takes into account some typical aspects of FANET, such as a human mobility model with the introduction of Points of Interest (PoIs) in order to simulate the main behavior of the people, the introduction of some typologies of users in order to simulate different users' moving, the footprint for channel modeling, the Link-State routing protocol for data exchange (audio and video streaming in particular), and dynamic range adjustment for taking into account user requests. Simulation results verified and confirmed the effectiveness of the proposed simulator.

As regards our future works, we plan to improve the simulator by adding some other aspects, such as other network protocols and various techniques for safety and security considering new proposal for the solutions against different attacks. It is also important to consider new optimal placement algorithms and investigate adequately the energy aspects for drone flight in every phase of its movement, including providing the connection to the users. Moreover, it is possible to improve the use of the recharging station through the optimization of the stations and the number of backup drones for each station. Moreover, in this work, no collision avoidance issues have been taken into account and no consideration has been done on the possibility of a drone to be in the same airspace of its neighbors. All these aspects are in the development phase and will be taken into account in our future proposals.

Author Contributions: F.D.R. conceived the idea; M.T. designed and implemented the software simulator with the help of P.F.; F.D.R. and N.C. analyzed the data; and M.T. and F.D.R. reviewed the paper with the help of N.C. All authors have read and agreed to the published version of the manuscript.

Funding: This research received no external funding.

Conflicts of Interest: The authors declare no conflict of interest.

References

1. Winkler, S.; Zeadally, S.; Evans K. Privacy and Civilian Drone Use: The Need for Further Regulation. *IEEE Secur. Priv.* **2018**, *16*, 72–80. [CrossRef]
2. Oubbati, O.S.; Atiquzzaman, M.; Lorenz, P.; Tareque, H.; Hossain S. Routing in Flying Ad Hoc Networks: Survey, Constraints, and Future Challenge Perspectives. *IEEE Access* **2019**, *7*, 81057–81105. [CrossRef]
3. Bujari, A.; Calafate, C.T.; Cano, J.C.; Manzoni, P.; Palazzi, C.E.; Ronzani, D. Flying ad-hoc network application scenarios and mobility models. *Int. J. Distrib. Sens. Netw.* **2017**, *13*, 10. [CrossRef]
4. Huang, H.; Savkin, A.V. An Algorithm of Efficient Proactive Placement of Autonomous Drones for Maximum Coverage in Cellular Networks. *IEEE Wirel. Commun. Lett.* **2018**, *7*, 994–997. [CrossRef]
5. Fazio, P.; Tropea, M.; De Rango, F.; Voznak, M. Pattern prediction and passive bandwidth management for hand-over optimization in QoS cellular networks with vehicular mobility. *IEEE Trans. Mob. Comput.* **2016**, *15*, 2809–2824. [CrossRef]
6. Fazio, P.; De Rango, F.; Tropea, M. Prediction and qos enhancement in new generation cellular networks with mobile hosts: A survey on different protocols and conventional/unconventional approaches. *IEEE Commun. Surv. Tutor.* **2017**, *19*, 1822–1841. [CrossRef]
7. Santamaria, A.F.; Fazio, P.; Raimondo, P.; Tropea, M.; De Rango, F. A New Distributed Predictive Congestion Aware Re-Routing Algorithm for CO_2 Emissions Reduction. *IEEE Trans. Veh. Technol.* **2019**, *68*, 4419–4433. [CrossRef]
8. Nguyen, T.N.; Duy, T.T.; Luu, G.T.; Tran, P.T.; Voznak, M. Energy harvesting-based spectrum access with incremental cooperation, relay selection and hardware noises. *Radioengineering* **2017**, *26*, 240–250. [CrossRef]
9. Nguyen, T.N.; Do, D.T.; Tran, P.T.; Voznak, M. Time switching for wireless communications with full-duplex relaying in imperfect CSI condition. *KSII Trans. Internet Inf. Syst.* **2016**, *10*, 4223–4239.

10. Do, D.T.; Nguyen, H.S.; Voznak, M.; Nguyen, T.S. Wireless powered relaying networks under imperfect channel state information: System performance and optimal policy for instantaneous rate. *Radioengineering* **2017**, *26*, 869–877. [CrossRef]

11. Tropea, M.; Fazio, P. A Simulator for Creating Drones Networks and Providing Users Connectivity. In Proceedings of the 2019 IEEE/ACM 23rd International Symposium on Distributed Simulation and Real Time Applications (DS-RT), Cosenza, Italy, 7–9 October 2019; pp. 1–8.

12. Stöcker, C.; Bennett, R.; Nex, F.; Gerke, M.; Zevenbergen, J. Review of the current state of UAV regulations. *Remote. Sens.* **2017**, *9*, 459. [CrossRef]

13. Bozza Circolare LIC 15 Mezzi Aerei a Pilotaggio Remoto - Centri di Addestramento e Attestati Pilota. Available online: https://www.enac.gov.it/la-normativa/normativa-enac/consultazione-normativa/bozza-circolare-lic-15-mezzi-aerei-pilotaggio-remoto-centri-di-addestramento-attestati-pilota (accessed on 10 November 2019).

14. Workshop "Prospettive e transizione dal Regolamento ENAC Mezzi a pilotaggio Remoto al nuovo Regolamento EASA". Available online: https://www.enac.gov.it/news/workshop-prospettive-transizione-dal-regolamento-enac-mezzi-pilotaggio-remoto-al-nuovo (accessed on 10 November 2019).

15. Mardani, A.; Chiaberge, M.; Giaccone, P. Communication-Aware UAV Path Planning. *IEEE Access* **2019**, *7*, 52609–52621. [CrossRef]

16. Song, X.; Hu, S. 2D path planning with Dubins-path-based A* algorithm for a fixed-wing UAV. In Proceedings of the 2017 3rd IEEE International Conference on Control Science and Systems Engineering (ICCSSE), Beijing, China, 17–19 August 2017.

17. Ghaddar, A.; Merei, A. Energy-Aware Grid Based Coverage Path Planning for UAVs. In Proceedings of the Thirteenth International Conference on Sensor Technologies and Applications SENSORCOMM, Nice, France, 27–31 October 2019.

18. De Rango, F.; Palmieri, N.; Santamaria, A.F.; Potrino, G. A simulator for UAVs management in agriculture domain. In Proceedings of the 2017 International Symposium on Performance Evaluation of Computer and Telecommunication Systems (SPECTS), Seattle, WA, USA, 9–12 July 2017; pp. 1–8.

19. De Rango, F.; Palmieri, N.; Tropea, M.; Potrino, G. UAVs Team and Its Application in Agriculture: A Simulati5on Environment. *SIMULTECH* **2017**, *2017*, 374–379.

20. De Rango, F.; Potrino, G.; Tropea, M.; Santamaria, A.F.; Fazio, P. Scalable and ligthway bio-inspired coordination protocol for FANET in precision agriculture applications. *Comput. Electr. Eng.* **2019**, *74*, 305–318. [CrossRef]

21. Marconato, E.A.; Rodrigues, M.; Pires, R.D.M.; Pigatto, D.F., Luiz Filho, C.Q.; Pinto, A.R.; Branco, K.R. Avens-a novel flying ad hoc network simulator with automatic code generation for unmanned aircraft system. In Proceedings of the HICSS, Hilton Waikoloa Village, HI, USA, 4–7 January 2017.

22. Zema, N.R.; Trotta, A.; Sanahuja, G.; Natalizio, E.; Di Felice, M.; Bononi, L. CUSCUS: CommUnicationS-control distributed simulator. In Proceedings of the 2017 14th IEEE Annual Consumer Communications & Networking Conference (CCNC), Las Vegas, NV, USA, 8–11 January 2017; pp. 601–602.

23. Baidya, S.; Shaikh, Z.; Levorato, M. FlyNetSim: An open source synchronized UAV network simulator based on ns-3 and ardupilot. In Proceedings of the 21st ACM International Conference on Modeling, Analysis and Simulation of Wireless and Mobile Systems, Montreal, QC, Canada, 19 October 2018; pp. 37–45.

24. Kate, B.; Waterman, J.; Dantu, K.; Welsh, M. Simbeeotic: A simulator and testbed for micro-aerial vehicle swarm experiments. In Proceedings of the 2012 ACM/IEEE 11th International Conference on Information Processing in Sensor Networks (IPSN), Beijing, China, 16–20 April 2012; pp. 49–60.

25. Javaid, A.Y.; Sun, W.; Alam, M. UAVSim: A simulation testbed for unmanned aerial vehicle network cyber security analysis. In Proceedings of the 2013 IEEE Globecom Workshops (GC Wkshps), Atlanta, GA, USA, 9–13 December 2013; pp. 1432–1436.

26. Al-Mousa, A.; Sababha, B.H.; Al-Madi, N.; Barghouthi, A.; Younisse, R. UTSim: A framework and simulator for UAV air traffic integration, control, and communication. *Int. J. Adv. Robot. Syst.* **2019**, *16*. [CrossRef]

27. Al-Hourani, A.; Kandeepan, S.; Jamalipour, A. Modeling air-to-ground path loss for low altitude platforms in urban environments. In Proceedings of the 2014 IEEE Global Communications Conference, Austin, TX, USA, 8–12 December 2014; pp. 2898–2904.

28. Park, K.N.; Cho, B.M.; Park, K.J.; Kim, H. Optimal coverage control for net-drone handover. In Proceedings of the 2015 Seventh International Conference on Ubiquitous and Future Networks, Sapporo, Japan, 7–10 July 2015; pp. 97–99.

29. Sae, J.; Yunas, S.F.; Lempiainen, J. Coverage aspects of temporary LAP network. In Proceedings of the 2016 12th Annual Conference on Wireless on-Demand Network Systems and Services (WONS), Cortina d'Ampezzo, Italy, 20–22 January 2016; pp. 1–4.

30. Xie, J.; Dong, C.; Li, A.; Wang, H.; Wang, W. Optimal Deployment Density for Maximum Coverage of Drone Small Cells. In Proceedings of the 2017 IEEE 86th Vehicular Technology Conference (VTC-Fall), Toronto, ON, Canada, 24–27 September 2017; pp. 1–6.

31. Papandrea, M.; Jahromi, K.K.; Zignani, M.; Gaito, S.; Giordano, S.; Rossi, G.P. On the properties of human mobility. *Comput. Commun.* **2016**, *87*, 19–36. [CrossRef]

32. Kang, J.H.; Welbourne, W.; Stewart, B.; Borriello, G. Extracting places from traces of locations. *ACM Sigmobile Mob. Comput. Commun. Rev.* **2005**, *9*, 58–68. [CrossRef]

33. Feng, Q.; McGeehan, J.; Tameh, E.K.; Nix, A.R. Path loss models for air-to-ground radio channels in urban environments. In Proceedings of the 2006 IEEE 63rd Vehicular Technology Conference, Melbourne, Victoria, Australia, 7–10 May 2006; Volume 6, pp. 2901–2905.

34. Fotouhi, A.; Qiang, H.; Ding, M.; Hassan, M.; Giordano, L.G.; Garcia-Rodriguez, A.; Yuan, J. Survey on UAV cellular communications: Practical aspects, standardization advancements, regulation, and security challenges. *IEEE Commun. Surv. Tutor.* **2019**, *21*, 3417–3442. [CrossRef]

35. Zeng, Y.; Zhang, R. Energy-efficient UAV communication with trajectory optimization. *IEEE Trans. Wirel. Commun.* **2017**, *16*, 3747–3760. [CrossRef]

36. Erdelj, M.; Natalizio, E. UAV-assisted disaster management: Applications and open issues. In Proceedings of the 2016 international conference on computing, networking and communications (ICNC), Kauai, HI, USA, 15–18 Febuary 2016; pp. 1–5.

Article

Onboard Visual Horizon Detection for Unmanned Aerial Systems with Programmable Logic

Antal Hiba [1], Levente Márk Sántha [2,*], Tamás Zsedrovits [2], Levente Hajder [3] and Akos Zarandy [1,2]

[1] Institute for Computer Science and Control (SZTAKI) Kende utca 13-17, 1111 Budapest, Hungary; hiba.antal@sztaki.hu (A.H.); zarandy.akos@sztaki.hu (A.Z.)

[2] Faculty of Information Technology and Bionics, Pazmany Peter Catholic University, Práter utca 50/A., 1083 Budapest, Hungary; zsedrovits.tamas@itk.ppke.hu

[3] Department of Algorithms and Their Applications, Eötvös Loránd University, Pázmány Péter stny. 1/C., 1117 Budapest, Hungary; hajder@inf.elte.hu

* Correspondence: santha.levente.mark@itk.ppke.hu; Tel.: +36-1-886-4700

Received: 15 March 2020; Accepted: 2 April 2020; Published: 4 April 2020

Abstract: We introduce and analyze a fast horizon detection algorithm with native radial distortion handling and its implementation on a low power field programmable gate array (FPGA) development board in this paper. The algorithm is suited for visual applications in an airborne environment, that is on board a small unmanned aircraft. The algorithm was designed to have low complexity because of the power consumption requirements. To keep the computational cost low, an initial guess for the horizon is used, which is provided by the attitude heading reference system of the aircraft. The camera model takes radial distortions into account, which is necessary for a wide-angle lens used in most applications. This paper presents formulae for distorted horizon lines and a gradient sampling-based resolution-independent single shot algorithm for finding a horizon with radial distortion without undistortion of the complete image. The implemented algorithm is part of our visual sense-and-avoid system, where it is used for the sky-ground separation, and the performance of the algorithm is tested on real flight data. The FPGA implementation of the horizon detection method makes it possible to add this efficient module to any FPGA-based vision system.

Keywords: UAS; UAV; horizon; undistortion; FPGA; sense-and-avoid

---

1. Introduction

Unmanned aerial vehicle systems (UAS) with an airborne camera are used in more and more applications from the aerial recreational photo shooting, to more complicated semi-autonomous surveillance missions, for example in precision agriculture [1]. Safe usage of these autonomous UAS requires Sense and Avoid (SAA) capability to reduce the risk of collision with obstacles and other aircraft. Surveillance mission setups include onboard cameras and a payload computer, which can also be used to perform SAA. Thus, computationally cheap camera-based solutions may need no extra hardware component. Most of the vision-based SAA methods use different approaches for intruders in sky and land backgrounds [2–4]. Thus, they can utilize horizon detection to produce fast sky segmentation. Attitude heading reference system (AHRS) is a compulsory module of UAS which provides Kalman-filtered attitude information from raw IMU (Inertial Measurement Unit) and GPS measurements [5–7]. This attitude information can enhance the sky-ground separation methods, because it can give an estimate of the horizon line in the camera image [8]; furthermore, if the horizon is a visible feature (planar scenes), it can also be also used to improve the quality of attitude information [9] or support visual serving for fixed-wing UAV landing [10]. Sea, large lakes, plains, and even a hilly environment at high altitudes give visible horizon in the camera view, which is worth detection onboard.

There are several solutions for horizon detection in the literature. In general, more sophisticated algorithms are used, which use a statistical model for the sky and non-sky region separation. Ettinger et al. use the covariance of pixels as a model for the sky and ground [11], while Todorovic uses a hidden Markov tree for the segmentation [12]. McGee et al. [13] and Boroujeni et al. [14] use a similar strategy, where the pixels are classified with a support vector machine, or with clustering. A different kind of strategy is followed by Cornall and Egan [15] and Dusha et al. [16] Cornall and Egan [15] use various textures, and they only calculate the roll angle. Dusha et al. [16] combine optical flow features with Hough transform and temporal line tracking, to estimate the horizon line. The problem with the single line model is that it cannot efficiently estimate the horizon in the case of tall buildings, or trees and high hills, where the horizon can be better modeled with more line segments. Shen et al. [17] introduced an algorithm for horizon detection on complex terrain. Pixel clustering methods, in general, have the potential to give sky-ground separation in any case, but they are computationally expensive. Notably, [18] presents a real-time solution that considers sky and ground pixels as fuzzy subsets in YUV color space and continuously trains a classifier above this representation and compares its output to precomputed binary codes of possible attitudes. This approach can be very effective if ground textures do not change fast during flight and the camera always sees the horizon. To overcome this problem, the authors used two 180-degree field of view (FOV) cameras. Their image representation was only 80×40 pixel, but the straight line horizon is a large feature and can be detected in such low resolution.

SAA and most of the main mission tasks need high-resolution images for the detection of small features and aim for high angular resolution at large FOV. With a simple down-sampling, one can provide input for existing horizon detection methods that work on low-resolution images, but we can reach higher accuracy in the original image, which has a fine angular resolution. We still need to do other manipulations in the high resolution for SAA and main mission tasks; thus, using this has no overhead. In our previous work [8], a simple intensity gradient sampling method was proposed, which fine-tunes an initial horizon line coming from the AHRS attitude. The computational need is only defined by the number of sample points and is independent from the image resolution. The single-shot approach is more stable, because detection errors cannot propagate to consecutive frames. The existence of a horizon can also be defined based on the AHRS estimate. This horizon detector was utilized in our successful SAA flight demonstration [2].

Large FOV optics have radial distortion, which does not degrade the detection of small features (intruder). However, it makes horizon detection challenging without complete undistortion of the image. In this paper, we introduce the advanced version of [8] with native radial distortion handling, in which only a few numbers of feature/sampling points need to be transformed, instead of the complete image. The mathematical representation of the distorted horizon line is given with the experimental approach to finding it. The method is evaluated on real flight test data. This paper presents the field programmable gate array (FPGA) implementation of the novel horizon detection module, with a theoretic complete vision system for SAA.

2. Horizon Detection Utilizing AHRS Data

The attitude heading reference system provides the Euler angles of the aircraft, which define ordered rotations around the reference North-East-Down coordinate system. First, it takes a rotation around Down (yaw), then a rotation around the rotated East (pitch), and finally, a rotation around the twice rotated North (roll), to have the actual attitude of the aircraft. We can also calibrate once the relative transformation of the camera coordinate system to the aircraft body coordinate system. With this information, one can define a horizon line in the image, which has calibration and AHRS errors. In this paper, we use this AHRS-based horizon as an initial estimation of the horizon line in the image.

2.1. Horizon Calculation from AHRS Data

Figure 1 presents the main coordinate systems and the method for horizon calculation form Euler angles. We define one point of the horizon in the image plane and the normal vector of it. To reach this, we transform the North (surface parallel) unit vector of world coordinate system to the camera coordinate system and get its intersection with the image plane, which will be one point of the horizon (it can be out of the borders), and we can also transform a unit vector pointing upwards in the world reference to have a normal vector as the projection of transformed upward vector to the image plane. The horizon line is given by a point P_0 and a normal vector $\vec{n}$ [8].

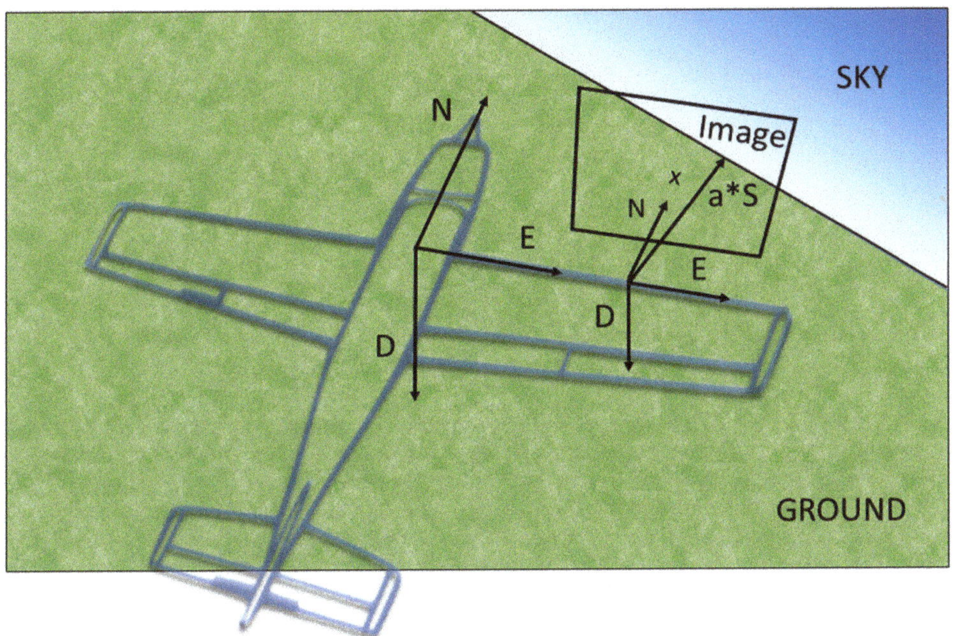

Figure 1. Blue aircraft with a camera on its wing. From the North-East-Down (NED) reference world coordinate system, we can derive the aircraft body frame NED (center of mass origin) and the camera frame NED (camera focal point origin) coordinate systems. S is a North unit vector in a word reference frame, which is parallel to the ground surface.

2.2. Horizon Based on AHRS Data Without Radial Distortion

The horizon is a straight line, if we assume a relatively unobstructed ground surface (flats, small hills, lakes, sea), with a negative intensity drop from the sky to ground, in most cases. Infrared cut-off filters can enhance this intensity drop, because it filters out infrared light reflected by plants. In general, because of the different textures on the images, the horizon cannot be found solely based on this intensity drop.

On the other hand, with AHRS data, an initial guess for the horizon line's whereabouts is available. This estimate is within a reasonable distance from the horizon in most cases. Thus, it can guide a search for the real horizon line on the image plane. Rotations around the center point of the line and shifts in the direction of its normal vector are applied to get new horizon candidates. Every time, the intensity drop is checked at given points, denoted by H (sky region) and L (ground region), as it is shown in Figure 2. The algorithm to find the horizon, in this case, is described in more detail in [8]. In this paper, we introduce the advanced version of this gradient sampling approach, which considers the radial distortion of large FOV cameras.

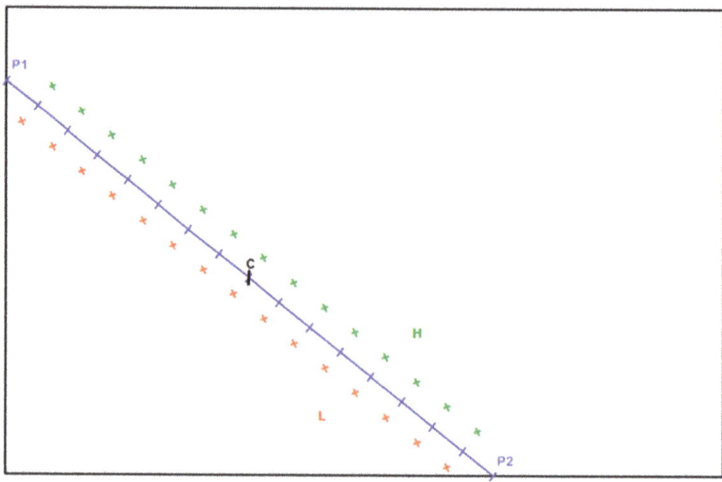

Figure 2. Horizon line candidate with the test point sets under (L—red) and above it (H—green).

2.3. Exact Formula of Distorted Horizon Lines

A standard commercial camera usually realizes the perspective projection. Consequently, the projection of the horizon is a straight line in the image space. However, perspectivity is only an approximation of the projection of real-world cameras. If a camera is relatively inexpensive or it has optics with a wide field of view (FOV), then more complex camera models should be introduced. A standard solution is to apply the radial distortion model to cope with the non-perspective behavior of cameras.

We have selected the 2-parameter radial distortion model [19] in this study. Using the model, the relationship between the theoretical perspective coordinates $[u \quad v]^T$ and the real ones $[u' \quad v']^T$ is given as follows:

$$\begin{bmatrix} u' \\ v' \end{bmatrix} = \left(1 + k_1 r^2 + k_2 r^4\right) \begin{bmatrix} u \\ v \end{bmatrix},$$

where k_1 and k_2 are the parameters of the radial distortion, and $r = \sqrt{u^2 + v^2}$ gives the distance between the point $[u \quad v]^T$, and the principal point. The principal point is the location at which the optical axis intersects the image plane. Therefore, this relationship is valid only if the origin of the used coordinate system is at the principal point. Another important pre-processing step is to normalize the scale of the axes by the division with the product of the horizontal and vertical focal length and pixel size of the camera.

The critical task for horizon detection is to compute the radially distorted variant of a straight line. Let the line be parameterized as:

$$\begin{bmatrix} u \\ v \end{bmatrix} = \begin{bmatrix} u_0 \\ v_0 \end{bmatrix} + t \begin{bmatrix} u \\ v \end{bmatrix},$$

where t is the line parameter, $d = [d_u \, d_v]$ the vector of the line's direction. The distorted line is written by:

$$\begin{bmatrix} u' \\ v' \end{bmatrix} = \left(1 + k_1 r^2 + k_2 r^4\right) \begin{bmatrix} u_0 + t d_u \\ v_0 + t d_v \end{bmatrix},$$

where: $r^2 = t^2\left(d_u^2 + d_v^2\right) + 2t(u_0 d_u + v_0 d_v) + u_0^2 + v_0^2$, and $r^4 = \left(r^2\right)^2$. The square of the radius can be written in a more simplified form as:

$$r^2 = At^2 + Bt + C,$$

where: $A = d_u^2 + d_v^2$, $B = 2(u_0 d_u + v_0 d_v)$, and $C = u_0^2 + v_0^2$. Then r^4 is expressed as:

$$r^4 = A^2 t^4 + 2ABt^3 + \left(B^2 + 2AC\right)t^2 + 2BCt + C^2,$$

Therefore, the formula $\left(1 + k_1 r^2 + k_2 r^4\right)$ is also a polynomial that is written in the following form:

$$\alpha t^4 + \beta t^3 + \gamma t^2 + \delta t + \epsilon,$$

where $\alpha = k_2 A^2$, $\beta = k_2 AB$, $\gamma = k_2\left(2AC + B^2\right) + k_1 A$, $\delta = 2k_2 BC + k_1 B$, and $\epsilon = k_2 C^2 + k_1 C + 1$. The parametric formula of the line can be written in a compact form as:

$$\left[\begin{array}{c} u\prime \\ v\prime \end{array}\right] = \left[\begin{array}{c} \sum_{i=0}^{5} a_i t^i \\ \sum_{i=0}^{5} b_i t^i \end{array}\right] \tag{1}$$

where: $a_5 = \alpha d_u$, $a_4 = \alpha u_0 + \beta d_u$, $a_3 = \beta u_0 + \gamma \beta d_u$, $a_2 = \gamma u_0 + \delta d_u$, $a_1 = \delta u_0 + \epsilon d_u$, and $a_0 = \epsilon u_0$. Similarly, $b_5 = \alpha d_v$, $b_4 = \alpha v_0 + \beta d_v$, $b_3 = \beta v_0 + \gamma \beta d_v$, $b_2 = \gamma v_0 + \delta d_v$, $b_1 = \delta v_0 + \epsilon d_v$, and $b_0 = \epsilon v_0$.

2.3.1. Limits

For the horizon detection, the radially distorted line must be sampled within the image. For this reason, the interval for parameter t must be determined in which the curve lies within the area of the image. Let us denote the borders of the image with u_l and u_r (horizontal), v_t and v_b (vertical). The corresponding values for parameter t are determined by solving the 5-degree polynomials $\sum_{i=1}^{5} a_i t^i - u_l = 0$, $\sum_{i=1}^{5} a_i t^i - u_r = 0$, $\sum_{i=1}^{5} b_i t^i - v_t = 0$ and $\sum_{i=1}^{5} b_i t^i - v_b = 0$. Each polynomial has five different roots. To our experiments, only one of those is the real root; the other four complex roots, two conjugate pairs, are obtained per polynomial. The complex roots are discarded, the minimal and maximal values of all possible real roots give the limits for parameter t.

2.3.2. Tangent Line

Another advantage of the formula defined in Equation (1). The direction $d_{\tan}$ of the tangent line can be determined trivially by deriving the equation. It is written as follows:

$$d_{\tan} = \left[\begin{array}{c} \sum_{i=0}^{5} i a_i t^{i-1} \\ \sum_{i=0}^{5} i b_i t^{i-1} \end{array}\right] \tag{2}$$

2.3.3. Example

It is demonstrated in Figure 3 that the distortion model and its formulae can be a good representation of the radially distorted variant of straight lines. The test image was taken by a GoPro Hero4 camera, containing wide-angle optics. This kind of lense usually produces visible radial distortion in the images. The camera was calibrated with images of a chessboard plane, using the widely-used Zhang calibration method implemented in OpenCV3 [20]. As is expected, the curves fit the real borders of the chessboard (left image) and the horizon (right). It is well visualized that the polynomial approximation is principally valid at the center of the image. There are fitting errors close to the border, as it is seen in the right image of Figure 3. This problem comes from the fact that the corners of the calibration chessboard cannot be detected near the borders. Therefore, calibration information is not available there.

Figure 3. Radially distorted lines in images taken by GoPro Hero4. Left: Distorted curves for borders of a chessboard plane. White color indicates the original straight line fractions, mostly covered by the blue, the corresponding radially distorted curves. Right: Curve of the horizon.

2.4. Horizon Detection, Based on AHRS Data with Radial Distortion

In the case of radial distortion, our straight-line approach [8] cannot be utilized, because the horizon's coordinate functions are transformed into a 5th-degree polynomial in the image space. One possible way is to undistort the whole image, and then the straight-line approach is viable. However, it takes several milliseconds, even with precomputed pixel maps. The previous section describes an exact method for the distorted horizon line calculation. However, it is not necessary to calculate the exact polynomial when the algorithm investigates a candidate horizon on a distorted image. Here, we present a method that approximates the distorted horizon line with a sequence of straight lines that connects the distorted sample points of a virtual straight horizon. Figure 4 shows the difference between an undistorted horizon line and the corresponding distorted curve.

Figure 4. The blue line represents the pre-calculated straight horizon with endpoints P1 and P2. Red crosses and the black curve represent the distorted version of the straight line.

The main steps of horizon detection are the same for the distorted and the distortion-free cases, as is summarized in Figure 5. The initial AHRS-based horizon is given the same way as in the undistorted case. The pre-calculated horizon is corrected based on the distorted image, with only a few and computationally very inexpensive modifications. The gradient sampling algorithm realizes the correction mechanism based on distorted visual input. The basic idea is to create sample points at the two sides of the straight version of a horizon candidate, and distort these points to get the necessary sampling coordinates in the image (alg. gradient sampling).

Algorithm Gradient Sampling: Horizon Post-calculation with Distortion

---

Require: AHRS horizon line (P_0, $\vec{n}$), num_of_samples, step_size, deg_range, shift_range, distort_func (k_1, k_2)

1. P_1, P_2 ← Two endpoints of the horizon line in the image, return if the horizon line is not on the image.
2.   *Center* ← $P_1 + (P_2 - P_1)/2$
3.   *Base_Points* ← num_of_samples number of equidistant sample points on the elongated AHRS horizon line
4.   **for** *deg* = deg_range.min **to** deg_range.max **do**
5.   *Rotated_Points*← Rotate *Base_Points* around the *Center* point by *deg*.
6. $\vec{n}_{local}$← Rotate $\vec{n}$ by *deg*
7.   **for** *shift_num* = shift_range .min **to** shift_range.max **do**
8. P_{local}←*Center* + *shift_num* * step_size * $\vec{n}_{local}$
9.   point set H ←*Rotated_Points* + (*shift_num* + 1) * step_size * $\vec{n}_{local}$
10.   point set L ← *Rotated_Points* + (*shift_num* − 1) * step_size * $\vec{n}_{local}$
11.   point set H ← distort_func (point set H)
12.   point set L ← distort_func (point set L)
13.   Calculate the average intensity difference at points H(i) and L(i)
14.   Update P_{best} and $\vec{n}_{best}$ searching for the maximal average difference
15.   **return** P_{best}, $\vec{n}_{best}$

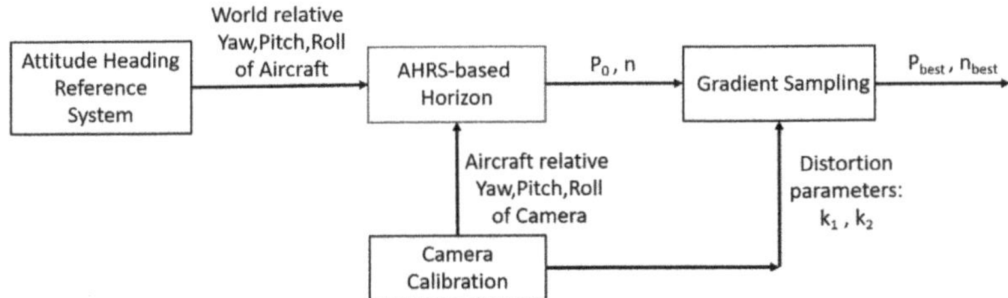

Figure 5. Block diagram of the horizon detection method. Camera calibration gives constants which should be defined only once. Gradient sampling method can consider radial distortion, with minimal overhead compared to the distortion-free version in [8].

The algorithm defines an exhaustive search of horizon line candidates. A predefined number of sample points are fixed on the initial pre-calculated horizon line (Base Points). These points are rotated around the center point (Rotated Points) and then shifted by the multiplicands of the rotated normal vector ($\vec{n}_{local}$), to get H and L sample point sets for each candidate line. Finally, the sample points are distorted to have proper sampling positions in the image, which has radial distortion. The average of $img(H(i)) - img(L(i))$ is calculated to give a gradient along with the horizon candidate. A predefined range of rotations and shifts is explored, and the line is chosen that has the largest average intensity difference on its two sides. If we consider the sample points of the straight horizon, and we connect the distorted versions of these points, it is possible that the resulted line series do not reach the borders of the image. Given that we want to get a complete horizon line, the pre-calculated straight horizon should be elongated. In our implementation, we use a rough solution by creating a 2*IMG WIDTH long line in all cases. Elements that are not in the image are discarded. Distortion and undistortion of pixel coordinates can be performed efficiently if we have a pre-computed Lookup table for the distortion of each image coordinate. Here, we need to remap only a predefined number of sample points instead of the whole image. The lines between the resulted sampling pairs (H-L) are not perfectly perpendicular to the corresponding horizon curve. However, this technique is still effective, because the L points are near below the horizon, and points in H are near above. Straight line-segments between sample points can define the horizon curve with a negligible difference from the real curve, as can be seen in Figures 6 and 7.

Figure 6. The blue dotted curve represents the pre-calculated horizon. Post-calculated horizon curve is formed by line segments between the green points. The effect of the white urban area under the hills can be seen on the right.

Figure 7. The blue dotted curve represents the pre-calculated horizon. The post-calculated horizon curve is formed by line segments between the green points. The small distortion error at the bottom corner can be seen on the left.

Sky-Ground Separation in Distorted Images

In our SAA application, the horizon line is used for sky and ground separation. In the case of straight-line horizons, the masking procedure is straightforward with the help of the normal vector. However, on the distorted image, we have a series of line segments. Furthermore, the ground mask may consist of two separate parts (the horizon curve goes out and then goes back to the image). To handle this problem, we use flood-fill operation started from two inner points next to the first, and the last points of the horizon curve on the image. The masks presented in the results section were generated this way.

3. Experimental Setup

The two variants of the horizon detection algorithm (with and without radial distortion) were tested in real flights. Flights were run at the Matyasfold public model airfield, which is close to Budapest, Hungary. There are only small hills around Matyasfold, resulting in a relatively straight horizon line, which was our original assumption.

3.1. Hardware Setup of Real Flight Tests

In the flight tests, a fixed-wing, two motor aircraft called Sindy was used. It is 1.85m in length, has a 3.5m wingspan, and has an approximately 12kg take-off weight (Figure 8). It is equipped with an IMU-GPS module, an onboard microcontroller with AHRS and autopilot functions (for details see [21]), and the visual sensor-processor system.

Figure 8. Sindy aircraft with the mounted Nvidia TK1-based visual SAA system.

We have two embedded GPU (Nvidia TK1, TX1) and two FPGA-based (Spartan 6 LX150T [22], Zynq UltraScale+ XCZU9EG) on-board vision system hardware. The development of new algorithms is much easier on a GPU platform. However, one can have the best power efficiency and parallelization with a custom FPGA implementation. All the flight test data in this paper acquired by the Nvidia Jetson TK1 system [8] and tested offline with the new FPGA system. The two Basler Dart 1280-54um cameras have monochrome 1280×960 sensor and 60-degree FOV optics, where the two FOVs have a 5-degree overlap.

There are different AHRS solutions which differ in sensor types and sensor fusion technique. Different levels of AHRS quality (with and without GPS sensor) and corresponding AHRS-based horizons were analyzed in [8]. In this paper, the best available on-board estimations of Euler angles are used to create AHRS-based horizon candidates. Our Kalman-filter based estimator is described in [6], which gives Euler angles to the autopilot. However, these results can still be improved, and small calibration errors of the camera relative attitude and deformations of the airframe during maneuvers can cause additional differences between the AHRS horizon and the visible feature in the image, which makes horizon detection necessary.

3.2. FPGA-Based Vision System Hardware

A field programmable gate array (FPGA)-based processing system is under development (Figure 9), on which it is possible to test finalized algorithms with more cameras at higher framerates, with lower power consumption compared to the embedded GPU. On the other hand, integration and test of such customized hardware to consider it flight-ready is very tedious, and we can also confirm its capabilities based on offline tests, with real flight data captured by the GPU-based system.

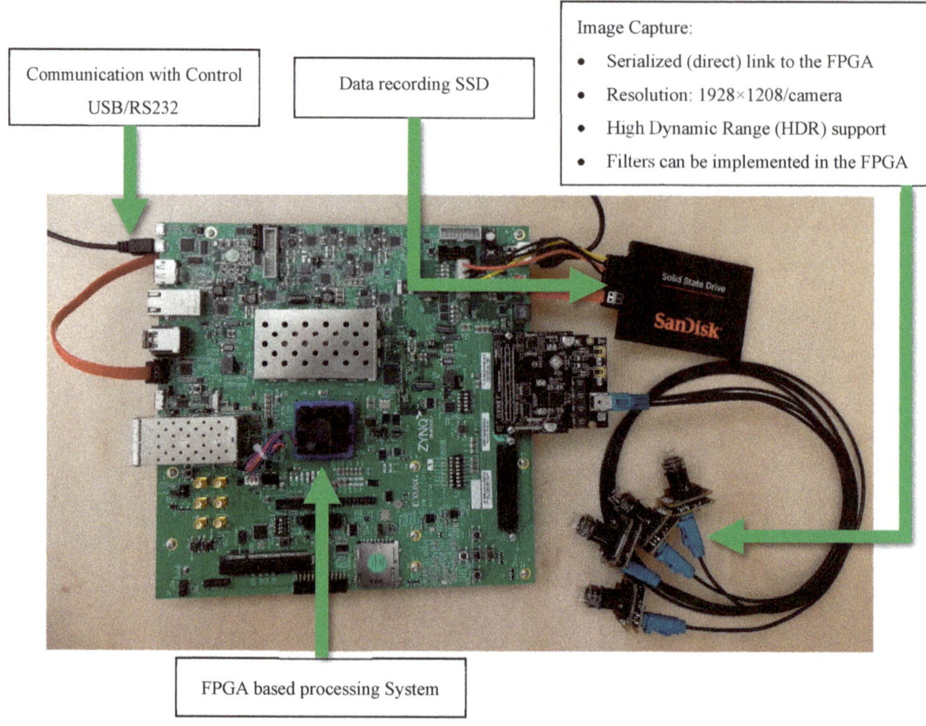

Figure 9. FPGA-based experimental system.

For research flexibility, we use a high-end FPGA Evaluation Board, Xilinx Zynq UltraScale+ Multi-Processor System-on-Chip (MPSoC) ZCU102. It contains various common industry-standard interfaces, such as USB, SATA, PCI-E, HDMI, DisplayPort, Ethernet, QSPI, CAN, I^{2}C, and UART.

For image capturing, an Avnet Quad AR0231AT Camera FMC Bundle set can be used. This contains an AES-FMC-MULTICAM4-G FMC module and four High Dynamic Range (HDR) camera modules, each with an AR0231AT CMOS image sensor (1928 × 1208), and MAX96705 serializer. Due to the flexibility of the system, this can be changed to other image capturing modules/methods, even to USB cameras.

4. FPGA Implementation

In this section, the FPGA implementation of the horizon detection module is presented for distortion-free and distorted images. This circuit is a part of an FPGA-based image processing system for collision avoidance. First, the whole (planned) architecture is briefly introduced, then the details of the realized horizon detection module are given with its power and programmable logic resource need on the FPGA.

4.1. Image Processing System on FPGA

The Zynq UltraScale+ XCZU9EG MPSoC FPGA chip has two main parts. The first is the FPGA Programmable Logic (PL). This contains the programmable circuit elements, such as look up tables (LUT), flip-flops (FF), configurable logic blocks (CLB), block memories (BRAM) and digital signal processing blocks (DSP) that are special arithmetic units designed to execute the most common operations in digital signal processing. The second part is called processing system (PS). Unlike the PL part, this contains fixed functional units, such as a traditional processor system, integrated I/Os and memory controller. PS has an application processing unit (APU) with four Arm Cortex-A53 cores at up to 1.5 GHz. A real-time processing unit (RPU) is also available with two Arm Cortex-R5 cores. There are fixed AXI BUS connections between the PL, PS, and the integrated I/Os.

Figure 10 shows the complete architecture of the SAA vision system. This architecture is based on [22] and not completely ready. In this paper, we introduce the horizon detection module and its role in the complete SAA system.

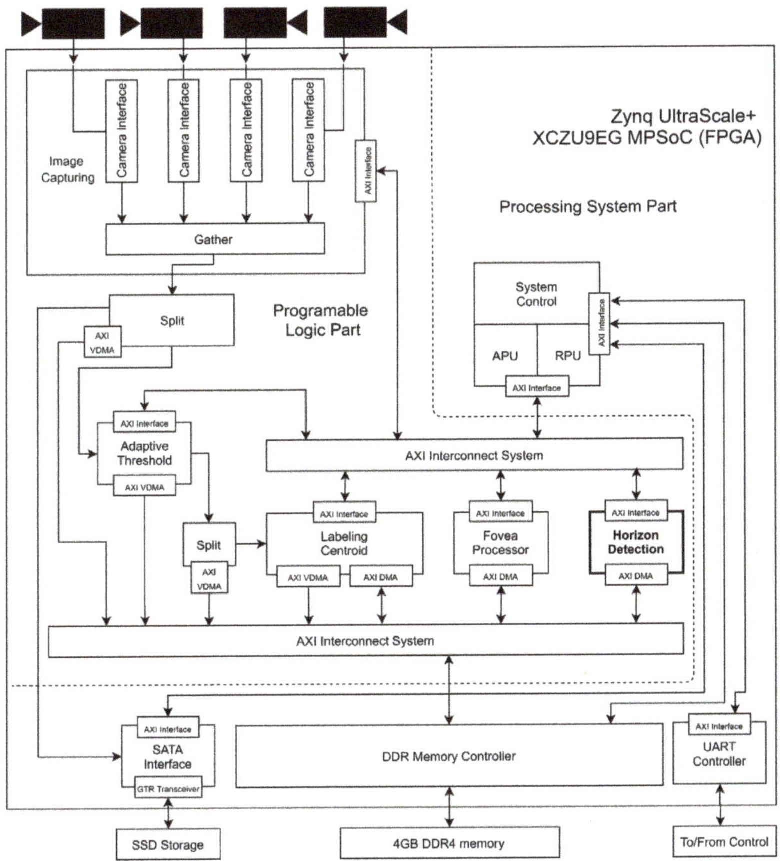

Figure 10. Block diagram of the FPGA based image processing system for SAA.

The PS part is used to control the image processing system. It sets the basic parameters of the different modules, reads their status information, and handles the communication with the control system. Some integrated I/Os are also utilized in this part. Raw image information is stored via the SATA interface. This can work as a black box during flight, and also makes it possible to use previously recorded flight data in offline tests. The DDR memory controller is connected to the onboard 4GB DDR4 memory. The UART controller performs the communication between the AHRS and the image processing system.

The computationally intensive part of the system is placed in the PL part. First, we capture the camera image of each camera and bind them together, making a new extended image. This is piped to three data paths: one to the SSD storage via the SATA interface, one to the memory (the other modules can access the raw image), and the third to the adaptive threshold module.

The input dataflow is examined by an n×n (now it's 5×5) window looking for high contrast regions, which are marked in a binary image, that is sent to the Labeling-Centroid module and stored in the memory. In the next step, objects are generated from the binary image. In the beginning, every marked region is an object and gets an identification label. If two marked regions are neighbors (to the left, right, up, down, or diagonally), they can be merged into one object. Due to some special shapes, like U, there is a second step in merging. The centroid of every object is calculated and stored.

When a full raw image is stored in the memory, the horizon detection module starts the calculation based on the AHRS data. The module returns the parameters of the horizon line, that can be used by the PS.

When both the horizon detection and the labeling-centroid module have finished, the PS is notified. Based on the current number of objects, the adaptive threshold module is set, as the number of objects on the next image should be between 20 and 30. Based on the horizon line, the PS determines that an object should be further investigated, or it can be dropped (the system detects sky objects only). The Fovea processor classifies the remaining objects, and the PS tracks the objects which were considered as intruders by the Fovea processor. The evasion maneuver is triggered based on the track analysis. For more details about the concept of this FPGA-based SAA image processing system, see [22].

4.2. Horizon Detection Module

In [8], the distortion-free version of the horizon detection was introduced and implemented on an Nvidia Jetson TK1 embedded GPU system. In this paper, we introduce direct handling of radial distortion without undistortion of the whole image, which is also implemented and used in-flight with the TK1 system. Here, we give the FPGA implementations for both the distorted and the distortion-free cases.

The TK1 C/C++ code is slightly modified to suit FPGA architecture design requirements better. High levels synthesis (HLS) tools were used to generate the hardware description language (HDL) sources from the modified C/C++ code. The main advantage of HLS compared to the more general pure HDL way is that the development time is much less, while it still gives low-level optimization possibilities.

4.2.1. Horizon Detection Without Radial Distortion

The block diagram of the AHRS-based pre-calculation [8] submodule is shown in Figure 11. First, the trigonometric functions of the Euler angles are calculated. The result is written in Rot1 or Rot2 memory, which works like a ping-pong buffer. In this implementation, when the result is available in one of the memories, the system starts the matrix multiplication. The transformation matrix from the body frame to the camera frame (BtoC) depends on only constants. Thus, it is calculated offline and stored in a ROM. The design is pipelined, therefore during the matrix multiplication, the system calculates another trigonometric function. When the *WC* matrix, which defines the projection from the world coordinate system to the camera frame, has been calculated, the system writes it to a memory module.

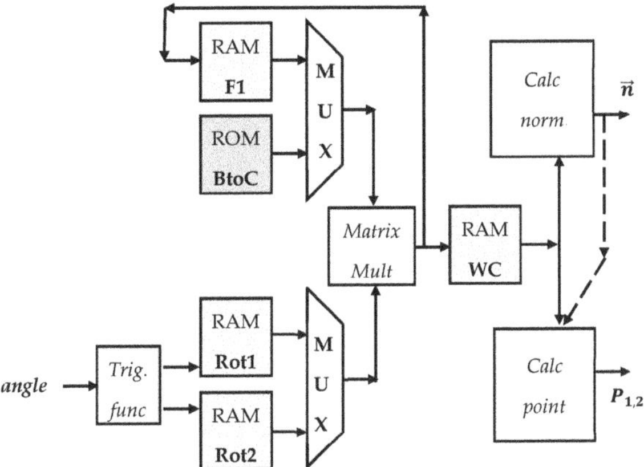

Figure 11. Pre-calculation block diagram. Blocks: F1: temporary memory for matrix multiplication. BtoC: Body frame to Camera Frame transformation matrix. Rot1, Rot2: Ping-pong buffer for trigonometric functions. WC: World to Camera frame transformation matrix.

After that, the transformation of the AHRS-horizon parameters, a normal vector (calc norm), and the two points of the line (calc point) occur in parallel. These two (three) values define the pre-calculated horizon line on the image.

The block diagram of the FPGA implementation of the post-calculation (Gradient Sampling) is shown in Figure 12. The endpoints of the pre-calculated horizon are derived from the normal vector and the point from the previous step. The "Case Calc" module is estimating the location of the endpoints on the image. For this estimation, eight regions are distinguished: The four sides and the four corners. This is used as an auxiliary variable for the calculations. It is necessary to check if the horizon line is in the image, based on the endpoints (check). If the endpoints are on the image, the coordinates of the base points (BP) are calculated, and these are stored in a RAM. Otherwise, the "valid" signal will be false, and the following steps are not calculated.

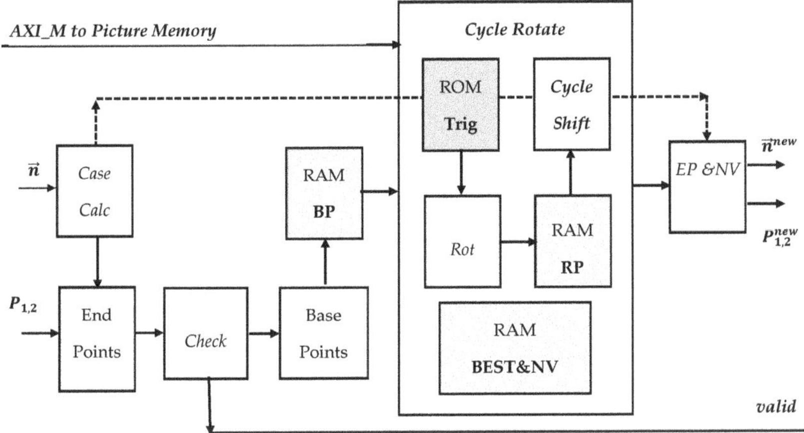

Figure 12. Post-calculation (gradient sampling) block diagram without radial distortion. Blocks: BP: Base point storage memory. Trig: Values of trigonometric functions used in rotations. Rot: Rotation calculation. RP: temporary storage for rotated points. BEST&NV: Temporarily stores the parameters of the best founded horizon line. EP&NV: Endpoints and normal vector calculation.

In a cycle of rotations, the base points are rotated in each step with a predefined angle. The values of the trigonometric functions (sin, cos) are calculated for these predefined angles, and they are stored in a ROM (trig). The result of the rotation is stored in RAM RP, and an inner cycle is run for the shift (cycle shift). In cycle shift, only those point pairs are used for the average intensity difference calculation, where both points are in the image. For each run, the result is compared to the best average so far, which is stored in RAM BEST&NV. The normal vector and a point of the line are also stored. The cycles are pipelined to speed up the calculation. In the end, the two endpoints and the normal vector of the horizon are calculated in EP&NV.

4.2.2. Horizon Detection with Radial Distortion

The pre-calculation submodule is nearly the same as in the distortion-free case. The only difference is that we calculate only one point; the center point, not two.

The post-calculation submodule required noticeable modifications. Due to the distortion, it can happen that we have even four intersections with the borders. Of course, in this case, only the left- and the rightmost intersections are used. Therefore, the case calculation block was eliminated, end the endpoint calculation module was extended with the functionality of handling the distortion. The other changes in this submodule were in the cycle shift block. The other functionalities of the blocks remained the same as the distortion-free case. In all shift cycles for every sample point pairs, the distortion is calculated, and the samples are taken from the distorted pixel coordinates. In the end, a normal vector and the center point of the (tuned) horizon are calculated in CP and NV block. The block diagram of the modified post-calculation submodule is shown in Figure 13.

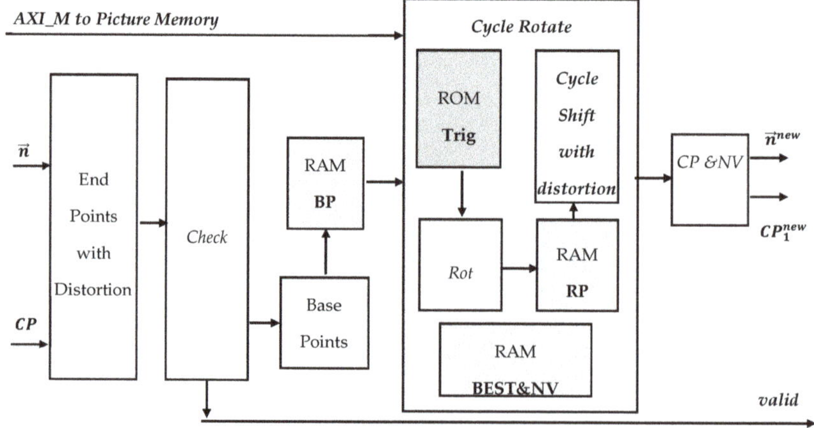

Figure 13. Post-calculation (gradient sampling) block diagram with radial distortion. Blocks: BP: Base point storage memory. Trig: Values of trigonometric functions used in rotations. Rot: Rotation calculation. RP: temporary storage for rotated points. BEST&NV: Temporarily stores the parameters of the best founded horizon line. CP&NV: Center point and normal vector calculation.

4.2.3. Resource and Power Usage

The two different FPGA modules (with and without radial distortion) were optimized by two different strategies. The first strategy aims for the highest possible clock rate/lowest evaluation time, but keeping the requirement that the total usage of the available resources cannot be greater than 20% (referred to as "Speed" implementation). The second one aims for less resource usage, even suitable for low-cost FPGA-s (referred to as "Area" implementation). Based on HLS technology, the C/C++ source during this process remained the same, and only the optimization directives/pragmas were changed. All implementations use floating-point number representations to have the same numeric properties as the TK1 implementation. However, these results can be further optimized with a fixed-point machine number representation. Execution time mainly depends on the number of sample points (20) of the gradient sampling phase and its search parameters, such as the number of rotation (±15 degree/ 1 degree step) and shift (±120 pixel/5 pixel step) steps.

The properties of each architecture are shown in Table 1. Even the resource optimized version (Area) with distortion has 2 mS execution time. The other versions have better results, but are relatively close to this. If we are comparing the speed and area versions, we can see that the resource usage can be reduced, especially when we calculate the distortion. On the other hand, the maximum clock frequency is also decreased. In this (area) case the resource elements (multipliers, adder, and others) are shared between the different blocks. This reduces the number of them, but generates a more complex control flow graph, thus a more complex final state machine that cannot operate at a high clock frequency as the speed implementation. Distorted and distortion-free versions do not have a huge difference in execution time. Distortion handling requires more circuit resources, four times more DSP, and two times more FF and LUT.

Table 1. The clock frequencies, execution times, and resource usage of different implementation strategies.

	Without Distortion		With Distortion	
Optimization Goal	Speed	Area	Speed	Area
Timing				
Clock Frequency	~110 MHz	~72 MHz	~100 MHz	~72 MHz
Execution Time	~1.2 mS	~1.8 mS	~1.4 mS	~2 mS
Resource				
BRAM_18K	2	2	2	2
DSP48E	70	48	211	75
FF	20,336	14,056	42,624	30,163
LUT	26,698	23,554	49,408	34,744
Resource %				
BRAM_18K	~0%	~0%	~0%	~0%
DSP48E	2%	1%	8%	4%
FF	3%	2%	7%	5%
LUT	9%	8%	18%	12%

The estimated power usage of the whole image processing system is below 5 Watts. It doesn't matter which horizon detection module is used, and it remains 5 W, because the most power consuming part of the system is the PS, not the PL in which the horizon module lies. More than 80% of the whole usage comes from the PS part. The estimation is based on the Xilinx Vitis/Vivado built-in power consumption estimator.

5. Experimental Results

The performance measurement of the horizon detection algorithm was done off-line, using the in-flight video and sensor data. Three flight video segments were analyzed; two of them consist of 1220 frames, and one of them consists of 1203 frames, which covers 2.5 min for each video considering the 8Hz sampling frequency. Each frame consists of two images; thus, altogether, more than 7000 horizons were evaluated. On each image, the horizon line was annotated by hand. Frames, where the horizon cannot be seen, are skipped from the calculations. There were around 1000 of these images. Tests for the undistorted case (2 videos) and the distorted case (1 video) were run on videos captured at Matyasfold.

The AHRS and the CAM results were tested against the annotation. Three error measures were calculated to show the performance of the horizon detection: (1) roll angle error (2) pitch angle error

(3) sky mask pixel ratio error. The geometric interpretation of these error terms is shown in Figures 14 and 15, and a graphical comparison of the algorithms to the annotation is shown in Figure 16.

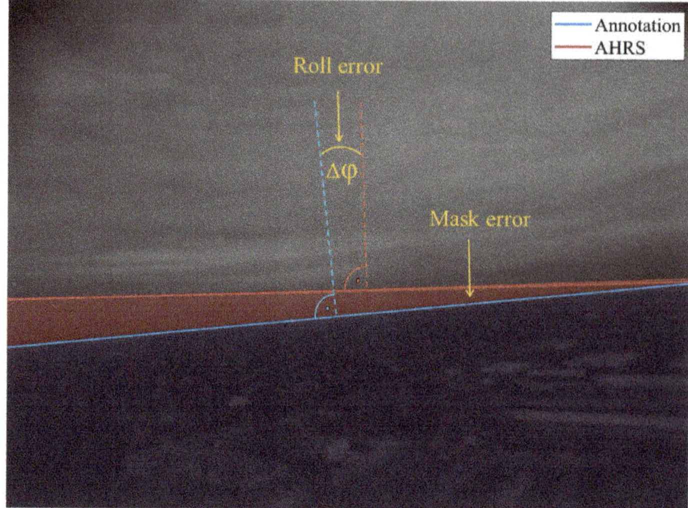

Figure 14. Roll and mask error definitions.

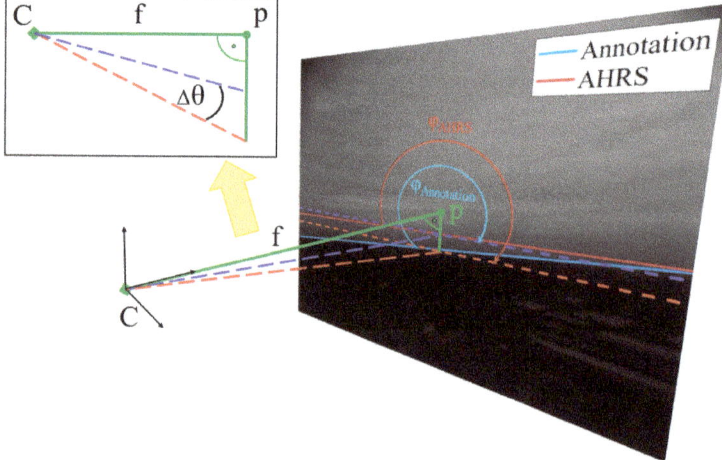

Figure 15. Pitch error definition. If the result is rotated around the image center by the roll, then the intersections of the horizons with the perpendicular line which goes through the center can give the pitch angle. We consider the vectors pointing from the 3D focal point of the camera.

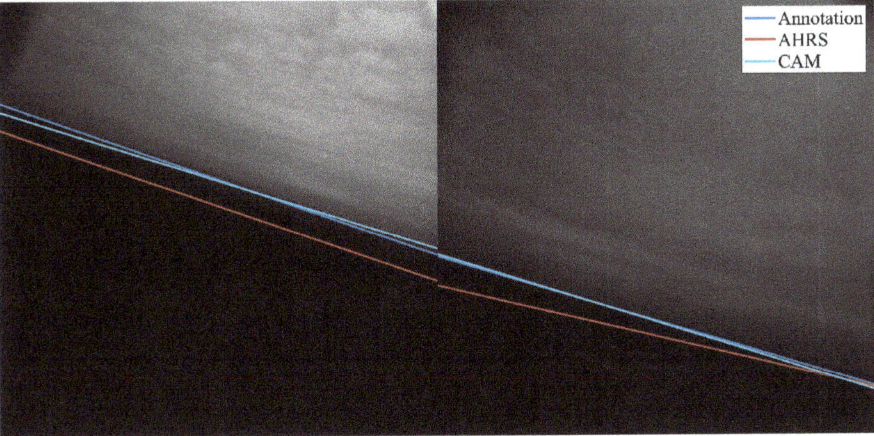

Figure 16. AHRS-based initial horizon candidate (red) the ground truth (blue) and the nearly identical final output of the gradient sampling (cyan).

AHRS of the Sindy aircraft with its full functionality (Inertial-Magnetic-Barometer—GPS) can provide attitude angles with less than 2-5-degree error. Thus, the horizon candidates of the pre-calculation step are close to the visual horizon in the image as it can be seen in Figure 16. The fine-tuned horizon output is nearly identical to the real visual horizon; only hills and long white urban structures near the horizon can disturb the fit. Figure 17 presents the effect of the gradient sampling phase, which can improve the AHRS horizon candidates from a 4.01 degree average absolute roll error to 1.73 degrees. We can see even better improvement for the pitch angle in Figure 18; from 6.75 degrees to 1.08 degrees.

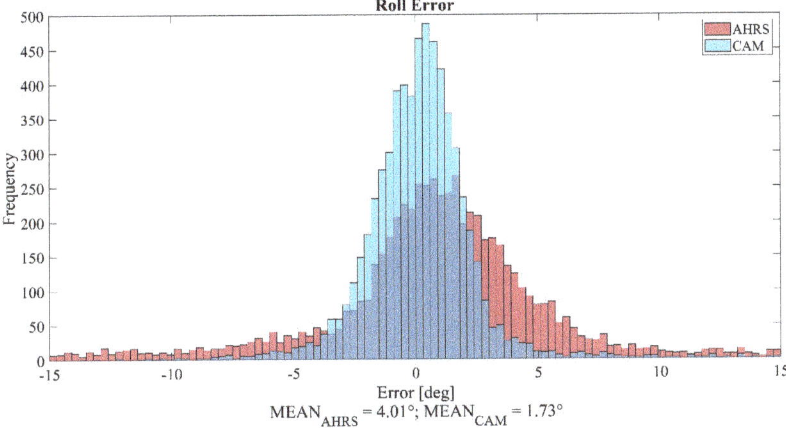

Figure 17. Roll error distributions of AHRS-only and the vision-based improvement on the real flight test data, with the average errors.

The bias of AHRS angles may come from the not ideal plane surface of the Mátyásföld area, and the possible small ~0.5-degree relative calibration error of the camera. With the elimination of this bias, we can still have benefited from the vision-based horizon, thanks to the fine angular resolution of our images compared to other methods that do not utilize the AHRS candidate and run complex sky-ground classification methods in low resolution to find a visual horizon. The real-time implementation of [18] reported a 1.49-degree average pith/roll error, which is close to other slower methods. Obviously, we have not solved the same challenging problem as other horizon detectors, which use only the image. However, we utilized the AHRS, which is always at hand on-board, and reach same performance on radially distorted images without undistortion of the complete image. The number of sample points (20), the number of rotation (+-15 degrees/1 degree resolution) and shift (+-120 pixel/5 pixel resolution) steps can be increased to reach better accuracy and robustness, however, the computational need has linear growth. This setup was suitable for our sky-ground separation application, where the main SAA mission task occupied most of our on-board computational resources.

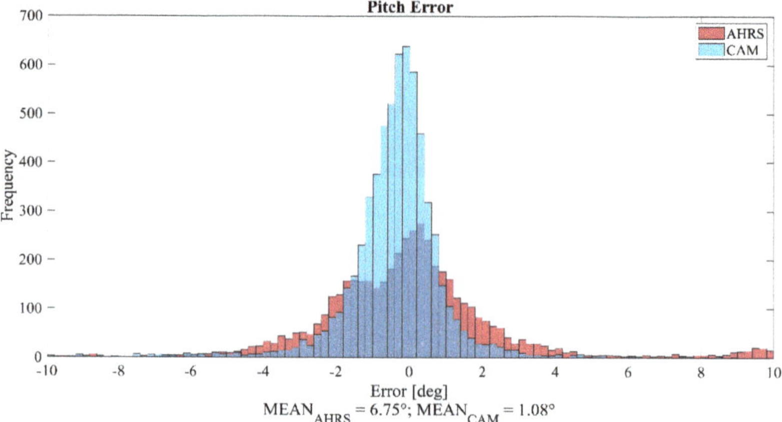

Figure 18. Roll error distributions of AHRS-only and the vision-based improvement on the real flight test data, with the average errors.

Due to the fact that we use the horizon for sky-ground separation, we also investigate the ground mask pixel error as the percentage of the image (Figure 19). Here, we can see the sum of roll and pitch improvement.

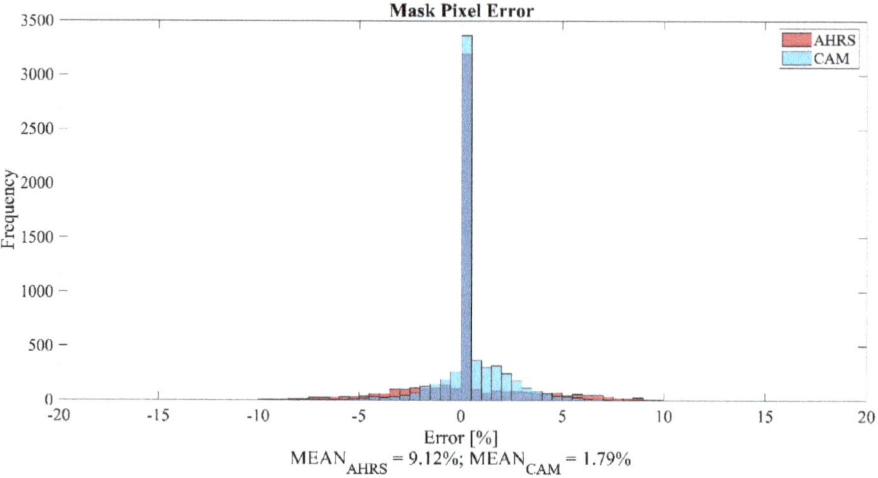

Figure 19. Pixel percentage error of ground masks for AHRS-only and the vision-based improvement on the real flight test data, with the average errors.

The gradient sampling with native distortion handling was initially implemented on the Nvidia Jetson TK1 embedded GPU and used in several flight tests with the Sindy aircraft. In this article, we gave the FPGA implementation for the horizon detection module to demonstrate that this problem can be solved under 5 W power in 2 mS time. Table 2 summarizes some reference algorithms and test platforms. Only [18] is close to the TK1 implementation. However, we have reached better accuracy, with the help of AHRS attitude information. The FPGA module for the distortion-free case can reach 1.2 mS response time with 20 sample points.

Table 2. The execution time of different horizon detection algorithms, on the given computer architecture and image size in the distortion-free case.

Algorithm	Computer Architecture	TDP	Camera Resolution (Resolution for Calculation)	Execution Time
Todorovic [12]	Athlon x86 @900 MHz	60 W	640 × 480 (128 × 128)	~600 mS
Ettinger et al. [11]	x86 @900 MHz	~60 W	320 × 240 (80 × 60)	~33 mS
McGee et al. [13]	Pentium III x86 @700 MHz	~24 W	320 × 240	~500 mS
Boroujeni et al. [14]	x86 @2.4 GHz (MATLAB)	~60 W	250 × 150	~2000 mS
Dusha et al. [16]	Pentium IV @3 GHz	84 W	352 × 288	~70 mS
Moore et al. [18]	dual-core PC104 @ 1.5 GHz	~20 W	360 × 180 (80 × 40)	~2 mS
Ours – NVIDIA TK1 [8]	ARM Cortex A15 @ 2.3 GHz	10 W	1280 × 960 (does not depend on resolution)	~5 mS
Ours – FPGA (Zynq UltraScale+ ZCU102)	Custom	<5 W	1928 × 1208 (does not depend on resolution)	~1.2 mS

Another way to handle distortion is to undistort the full image and then, run the simple horizon algorithm. In the literature, most of the undistortion algorithms consider a region of interest (ROI) and only distort this smaller area instead of the full image [23], which makes it possible to use simpler distortion models in a small part of the scene. In our case, we cannot predefine a ROI. Therefore, these methods are not suitable. In general, the system latency is increased with the full image undistortion time, because the horizon search algorithm can be started only with the fully undistorted image. Even if a fix-point number-representation is used, it takes around 20 mS on a full HD image (based on [24]). In our solution, the difference between the distorted and the undistorted version is only 0.2 mS (more FPGA circuit resources are consumed).

6. Conclusions

This paper presents a light-weight gradient sampling-based visual horizon detection for radially distorted images. Operating in planar environments where the horizon is a visible feature, the method explores the neighborhood of a horizon candidate, which is defined by the yaw pitch roll angles of the attitude heading reference system (AHRS), which is a piece of common equipment on a UAS. Vision-only methods are computationally more expensive, thus they are performed in low resolution and require radial undistortion of the image. In our case, the complete undistortion of the image is not necessary; only a few numbers of sample points should be transformed. The computational complexity of the method is independent from the image size, thus down sampling is not necessary. The original fine resolution image can provide better angular accuracy for the horizon detection.

The exact form of the distorted horizon is derived, which is a fifth order polynomial, and makes it possible to have the best quality representation if it is needed. The gradient sampling algorithm can efficiently improve even good quality AHRS data, based on the visual horizon. FPGA implementation is also given for the horizon detection, which demonstrates that the distortion handling can be performed with minimal time need (1.2 ms), if we add some extra circuit resources to the module. This realization is the fastest known vision-based horizon detector. The time complexity is defined by the number of sample points, rotation and shift steps. Our on-board setup has 1.7 degree roll and 1 degree pitch average absolute error, which is suitable for most applications and better than the vision-only real-time methods.

If complete undistortion of the whole image is not necessary for the main mission of the UAV, the horizon detection does not require it. A simple search around the AHRS-based horizon can reach high-quality attitude or sky-ground image masks.

Author Contributions: Conceptualization, A.H., A.Z., L.H. and T.Z.; data curation, A.H.; formal analysis, A.H., L.H. and T.Z.; investigation A.H., L.M.S. and T.Z.; methodology: A.H., L.H., L.M.S., A.Z. and T.Z.; software: A.H., T.Z. and L.M.S.; validation: A.H. and T.Z.; visualization: L.H., A.H., L.M.S. and T.Z.; writing—original draft: A.H., L.H., L.M.S. and T.Z.; writing—review and editing: A.Z., L.M.S., A.H. and T.Z. All authors have read and agree to the published version of the manuscript.

Funding: This research was funded by ONR, grant number N62909-10-1-7081. The APC was funded by PPCU KAP 19. L. Hajder was supported by the Project no. *ED_18-1-2019-0030* (Application-specific highly reliable IT solutions). The project has been implemented with the support provided from the National Research, Development and Innovation Fund of Hungary, financed under the Thematic Excellence Programme funding scheme. The article publication was funded by PPCU supported by NKFIH, financed under Thematic Excellence Programme.

Acknowledgments: Authors thank Krisztina Zsedrovitsne Gocze for the annotation of many flight videos, and Peter Bauer for the Sindy aircraft AHRS integration and flight tests.

Conflicts of Interest: The authors declare no conflict of interest.

References

1. Mogili, U.R.; Deepak, B.B.V.L. Review on application of drone systems in precision agriculture. *Procedia Comput. Sci.* **2018**, *133*, 502–509. [CrossRef]
2. Zsedrovits, T.; Peter, P.; Bauer, P.; Pencz, B.J.M.; Hiba, A.; Gozse, I.; Kisantal, M.; Nemeth, M.; Nagy, Z.; Vanek, B.; et al. Onboard visual sense and avoid system for small aircraft. *IEEE Aerosp. Electron. Syst. Mag.* **2016**, *31*, 18–27. [CrossRef]
3. Fasano, G.; Accardo, D.; Tirri, A.E.; Moccia, A.; De Lellis, E.E. Morphological filtering and target tracking for vision-based UAS sense and avoid. In Proceedings of the 2014 International Conference on Unmanned Aircraft Systems (ICUAS), Orlando, FL, USA, 27–30 May 2014; pp. 430–440.
4. Molloy, T.L.; Ford, J.J.; Mejias, L. Detection of aircraft below the horizon for vision-based detect and avoid in unmanned aircraft systems. *J. Field Robot.* **2017**, *34*, 1378–1391. [CrossRef]
5. Gleason, S.; Gebre-Egziabher, D. *GNSS Applications and Methods*; Artech House: Washington, DC, USA, 2009.
6. Bauer, P.; Bokor, J. Multi-mode extended Kalman filter for aircraft attitude estimation. *IFAC Proc. Vol.* **2011**, *44*, 7244–7249. [CrossRef]
7. Cornall, T.D.; Egan, G.K. Measuring Horizon Angle from Video on a Small Unmanned Air Vehicle. *Auton. Robots* **2004**, 339–344.
8. Hiba, A.; Zsedrovits, T.; Bauer, P.; Zarandy, A. Fast horizon detection for airborne visual systems. In Proceedings of the 2016 International Conference on Unmanned Aircraft Systems (ICUAS), Arlington, VA, USA, 7–10 June 2016; pp. 886–891.
9. Zsedrovits, T.; Bauer, P.; Hiba, A.; Nemeth, M.; Pencz, B.J.M.; Zarandy, A.; Vanek, B.; Bokor, J. Performance Analysis of Camera Rotation Estimation Algorithms in Multi-Sensor Fusion for Unmanned Aircraft Attitude Estimation. *J. Intell. Robot. Syst.* **2016**, *84*, 759–777. [CrossRef]
10. Gibert, V.; Burlion, L.; Chriette, A.; Boada, J.; Plestan, F. Nonlinear observers in vision system: Application to civil aircraft landing. In Proceedings of the 2015 European Control Conference (ECC), Linz, Austria, 15–17 July 2015; pp. 1818–1823.
11. Ettinger, S.M.; Nechyba, M.C.; Ifju, P.G.; Waszak, M. Towards flight autonomy: Vision-based horizon detection for micro air vehicles. In Proceedings of the Florida Conference on Recent Advances in Robotics, Melbourne, FL, USA, 14–16 May 2002.

12. Todorovic, S. Statistical Modeling and Segmentation of Sky/Ground Images. Master's Thesis, University of Florida, Gainesville, FL, USA, 2002.

13. McGee, T.G.; Sengupta, R.; Hedrick, K. Obstacle Detection for Small Autonomous Aircraft Using Sky Segmentation. In Proceedings of the 2005 IEEE International Conference on Robotics and Automation, Barcelona, Spain, 18–22 April 2005; pp. 4679–4684.

14. Boroujeni, N.S.; Etemad, S.A.; Whitehead, A. Robust horizon detection using segmentation for UAV applications. In Proceedings of the 9th Conference on Computer and Robot Vision, Toronto, ON, Canada, 28–30 May 2012; pp. 346–352.

15. Bauer, P.; Bokor, J. Development and hardware-in-the-loop testing of an Extended Kalman Filter for attitude estimation. In Proceedings of the 11th IEEE International Symposium on Computational Intelligence and Informatics (CINTI), Budapest, Hungary, 18–20 November 2010; pp. 57–62.

16. Dusha, D.; Boles, W.; Walker, R. *Fixed-Wing Attitude Estimation Using Computer Vision Based Horizon Detection*; QUT ePrints: Queensland, Australia, 2007; pp. 1–19.

17. Shen, Y.F.; Krusienski, D.; Li, J.; Rahman, Z. A Hierarchical Horizon Detection Algorithm. *IEEE Geosci. Remote Sens. Lett.* **2013**, *10*, 111–114. [CrossRef]

18. Moore, R.J.; Thurrowgood, S.; Bland, D.; Soccol, D.; Srinivasan, M.V. A fast and adaptive method for estimating UAV attitude from the visual horizon. In Proceedings of the 2011 IEEE/RSJ International Conference on Intelligent Robots and Systems, San Francisco, CA, USA, 25–30 September 2011; pp. 4935–4940.

19. Hartley, R.; Zisserman, A. *Multiple View Geometry in Computer Vision*; Cambridge University Press: Cambridge, UK, 2004; ISBN 0521540518.

20. Zhang, Z. A flexible new technique for camera calibration. *IEEE Trans. Pattern Anal. Mach. Intell.* **2000**, *22*, 1330–1334. [CrossRef]

21. Vanek, B.; Bauer, P.; Gozse, I.; Lukatsi, M.; Reti, I.; Bokor, J. Safety Critical Platform for Mini UAS Insertion into the Common Airspace. In Proceedings of the AIAA Guidance, Navigation, and Control Conference, American Institute of Aeronautics and Astronautics, Reston, VA, USA, 13–17 January 2014; pp. 1–13.

22. Zarándy, A.; Nemeth, M.; Nagy, Z.; Kiss, A.; Sántha, L.M.; Zsedrovits, T. A real-time multi-camera vision system for UAV collision warning and navigation. *J. Real-Time Image Process.* **2016**, *12*, 709–724. [CrossRef]

23. Jakub, C.; Henryk, B.; Kamil, G.; Przemysław, S. A Fisheye Distortion Correction Algorithm Optimized for Hardware Implementations. In Proceedings of the 21st International Conference "Mixed Design of Integrated Circuits & Systems", Lublin, Poland, 19–21 June 2014.

24. Daloukas, K.; Antonopoulos, C.; Bellas, N.; Chai, S. Fisheye lens distortion correction on multicore and hardware accelerator platforms. In Proceedings of the IEEE International Symposium on Parallel Distributed Processing, Atlanta, GA, USA, 19–23 April 2010; pp. 1–10.

 *electronics*

Article

LoRaWAN Networking in Mobile Scenarios Using a WiFi Mesh of UAV Gateways

Marco Stellin [1,2,3,†], Sérgio Sabino [1,2,4,†] and António Grilo [1,2,4,*,†]

[1] Instituto Superior Técnico, Universidade de Lisboa, Av. Rovisco Pais, no. 1, 1649-004 Lisbon, Portugal;
     stellin.marco@gmail.com (M.S.); sergio.sabino@tecnico.ulisboa.pt (S.S.)
[2] INESC-ID, Rua Alves Redol, no. 9, 1049-001 Lisbon, Portugal
[3] MobileKnowledge, Carrer de Roc Boronat, 117, 08018 Barcelona, Spain
[4] INOV, Rua Alves Redol, no. 9, 1049-001 Lisbon, Portugal
*   Correspondence: antonio.grilo@inesc-id.pt; Tel.: +351-213100226
†   These authors contributed equally to this work.

Received: 15 March 2020; Accepted: 5 April 2020; Published: 10 April 2020

Abstract: Immediately after a disaster, such as a flood, wildfire or earthquake, networks might be congested or disrupted and not suitable for supporting the traffic generated by rescuers. In these situations, the use of a traditional fixed-gateway approach would not be effective due to the mobility of the rescuers. In the present work, a double-layer network system named LoRaUAV has been designed and evaluated with the purpose of finding a solution to the aforementioned issues. LoRaUAV is based on a WiFi ad hoc network of Unmanned Aerial Vehicle (UAV) gateways acting as relays for the traffic generated between mobile LoRaWAN nodes and a remote Base Station (BS). The core of the system is a completely distributed mobility algorithm based on virtual spring forces that periodically updates the UAV topology to adapt to the movement of ground nodes. LoRaUAV has been successfully implemented in ns-3 and its performance has been comparatively evaluated in wild area firefighting scenarios, using Packet Reception Ratio (PRR) and end-to-end delay as the main performance metrics. It is observed that the Connection Recovery and Maintenance (CRM) and Movement Prediction (MP) mechanisms implemented in LoRaUAV effectively help improve the PRR, with the only disadvantage of a higher delay affecting a small percentage of packets caused by buffer delays and disconnections.

Keywords: LoRaWAN; Unmanned Aerial Vehicles; topology control; virtual spring forces; firefighting communications

---

1. Introduction

Over the past few years, the ubiquity of the Internet and the miniaturization of computational devices created a new paradigm called Internet of Things (IoT). In many cases, IoT devices are subject to very strict power constraints. For this reason, a new range of low power wireless communication protocols has been developed in order to support Low Power Wide Area Networks (LPWANs) [1]. These networks are formed by simple devices that communicate infrequently over long distances at low bitrates. Long Range (LoRa) is one of the most promising and versatile technologies enabling LPWANs. LoRa [2] is a narrowband modulation technique based on Chirp Spread Spectrum (CSS) modulation, a technology that achieves high robustness against channel degradation factors, such as path loss, multipath fading, shadowing and Doppler shift. By taking advantage of Spread Spectrum (SS), chirp orthogonality and the good propagation properties of the sub-GHz spectrum, LoRa provides communication over long distances, at the expense of the bitrate and of the maximum time interval between transmissions due to duty cycle limitations in the bands used by the protocol. Air time, power consumption and data rates can be controlled by different Spreading Factors (SFs) (7 to 12) and bandwidth (125 kHz or 250 kHz) combinations. The LoRa modulation is the core of the PHY and MAC of LoRa Wide Area Networks Protocol (LoRaWAN) [3].

Unmanned Aerial Vehicles (UAVs), popularly called drones, are flying vehicles that work without a human pilot onboard. The exclusive prerogative of the military for many years, UAVs are now commercially available at low prices, thus making them appealing for a wide variety of applications. Drones come in different shapes, but the most common are quadrotors and fixed-wing drones. In case of a disaster, such as a flood or a wildfire, UAVs can be used to support the rescuers, to localize the victims and to create a backhaul network when communication facilities are disrupted. UAVs provide many advantages compared to other solutions: movement in an obstacle-free environment, better overview of the area, Line of Sight (LoS) with targets and faster data acquisition in large areas. However, UAVs still have a limited flying range, autonomy and flying time. UAV swarms are increasingly being considered as a possible solution to provide radio coverage in a target area. Integrating UAVs and LPWAN protocols in disaster scenarios may offer a new cost-effective and energy efficient way to solve problems arising during the operations of the rescuers. In this paper, a challenging scenario was selected: support of wildfire combat operations. Wildfires affect rural or suburban areas where the coverage of conventional networks (e.g., cellular networks) is weak or absent, or vulnerable to the fire itself. UAVs can therefore be used to establish a relay network between the command post managing the operations and the firefighters, therefore providing situational awareness to the rescuers.

The mobility of firefighters represents a big challenge, since the UAV swarm has to maintain both area coverage and end-to-end connectivity to the Base Station (BS) of the command post. LoRaUAV, an UAV system based on LoRaWAN and Wireless Fidelity (WiFi), is designed to tackle these problems. Figure 1 shows a representation of the system architecture.

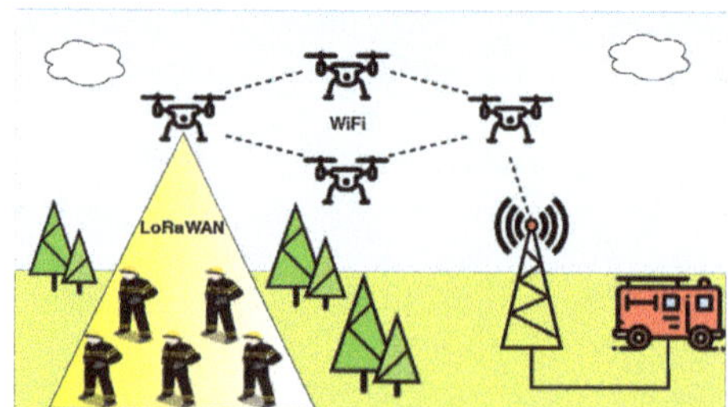

Figure 1. High level view of LoRaUAVsystem architecture.

In LoRaUAV, firefighters carry Global Positioning System (GPS) enabled LoRaWAN tags, which are used to transmit their position, as well as biometric data. They may also deploy tactical LoRaWAN sensors, which may detect when the fire has reached some selected locations. Firefighter tags and ground sensors will be henceforth designated Ground Nodes (GNs).

In the target scenario, the LoRaWAN radio range is likely affected by additional attenuation due to terrain features and foliage. In the LoRaUAV system, a WiFi mesh network of autonomous UAV LoRaWAN gateways is dynamically deployed to provide LoRaWAN coverage during firefighting operations. This mesh of flying LoRaWAN gateways will receive the data from the GNs and relay it through WiFi to the command post, where the LoRaWAN Network Server resides. Since the WiFi technology presents a higher data rate, it allows the aggregation of traffic originated from a high number of GNs. On the other hand, its shorter range is somewhat compensated by the fact that communication between UAVs takes place in LoS.

Since GNs are expected to move, the UAV mesh must adapt its position and topology according to the GN movement patterns. In LoRaUAV, this is addressed by a topology control algorithm based on Virtual Spring Forces (VSFs). In order to solve the problem of disconnections due to disruption caused by continuous movement of GNs and transient UAV topology adaptation, a Connection Recovery and Maintenance (CRM) extension is proposed. A Movement Prediction (MP) extension is also proposed in order to make it easier to recover isolated GNs based on a combination of movement prediction and VSFs.

LoRaUAV has been successfully implemented in ns-3and its performance has been evaluated and compared with an existing VSF proposal by Di Felice et al. [4]. Packet Reception Ratio (PRR) and end-to-end delay were used as the main performance metrics. From the results, it can be concluded that the CRM and MP mechanisms implemented in LoRaUAV effectively help improving the PRR, with the only disadvantage of a higher delay affecting a small percentage of packets, which is caused by buffer delays and disconnections.

The main contributions of this paper are the following.

- Proposal and comparative performance evaluation of a system providing LoRaWAN coverage in wild area firefighting operations by means of a WiFi mesh of UAV gateways.
- Novel CRM and MP extensions to VSFs topology control, which are shown to significantly improve the performance of this kind of mechanism.
- Novel mobility model of wild area firefighting scenarios considered in the performance evaluation.

The rest of this paper is organized as follows. Section 2 presents a review of the related works. Section 3 presents LoRaUAV and its main algorithms and explains the main design choices. Section 4 presents the developed simulation model. Section 5 presents the simulation results to evaluate all the algorithms. Finally, conclusions are reported in Section 6.

2. Related Work

Establishing a relay wireless ad hoc network of UAVs comes with a set of non trivial problems: how to optimally place the UAVs in a 3D environment, and how to plan the movement of UAVs to avoid collisions while reaching the target objective? A problem that frequently arises is the connectivity versus coverage problem. In an ideal situation, the mesh network should provide the maximum possible coverage, while at the same time it should maintain the connectivity between its members. Unfortunately, due to the limited available resources (i.e., number of UAVs), a compromise between both requirements is typically needed. The algorithms that have been developed to tackle the aforementioned problems fall in two main categories: centralized and distributed. Centralized algorithms rely on one single entity having full knowledge and control of the nodes forming the network. This approach typically produces close to optimal results when the location information is timely and reliable, but suffers from the single point of failure problem and, if the dimension of the problem is big enough, location information dissemination and the amount of computations can generate substantial delays. Mixed integer programming [5], evolutionary algorithms [6,7] or potential fields [8] are typically used. On the other hand, distributed algorithms provide less optimal solutions, but computations are typically simpler, based on local information distributed among the nodes, thus making the network more responsive and resilient in case of unexpected changes. Most of the existing distributed algorithms adapt concepts coming from physics or natural animal behavior. Such is the case of VSF approaches like the one presented in this paper. Consequently, this section will focus on distributed approaches proposed so far.

Basu et al. [9] propose a flocking based system, where UAVs self-organize to follow military units deployed and moving on the ground. The main purpose is to minimize the number of allocated UAVs in comparison with a full area coverage scheme. Fixed wing UAVs are assumed, with a constraint being considered on the maximum turn angle. A state machine with four states is proposed, which runs independently in each UAV, leading to an emergent flocking behavior. The state machine defines when the UAV should be repelled or attracted by its neighbors depending on mutual distance, when to approach the centroid of ground nodes within range, or when to randomly walk to find positions where more ground nodes are covered. The results show that the UAVs are able to follow and effectively cover the ground nodes when they move, though performance is clearly better when ground node movement is local and takes place within a constrained area.

Goddemeier et al. [10] consider a distributed decision approach to maintain a coherent mesh network of UAVs with the objective of exploring a 3D area. The swarm has also the additional requirement of keeping the connectivity to a ground base station. Two scenarios are studied by the authors: one in which the connectivity to the base station is permanent (Bounded Relaying) and one in which some disconnections are allowed for the purpose of extending the exploration (Release and Return). Each UAV in the network has the ability of self-selecting a different role, depending on the topology of the network: Scout Agents (SA) are assigned the task of exploration and sensing, Relay Nodes (RN) keep the communication between one or more UAVs and the base station, Articulation Points (AP) link two clusters that otherwise would be fragmented and Returnees (R) are drones that, in a Release and Return scenario, return to regain connectivity after being detached from the network. In order to maintain mesh connectivity, the authors propose a communication-aware algorithm based on virtual potential fields called Communication Aware Potential Fields (CAPF). According to this algorithm, each UAV is subject to a virtual force

$$\overrightarrow{F}_{dir} = \overrightarrow{F}_{conn} + \overrightarrow{F}_{AP} + \overrightarrow{F}_{CA}, \tag{1}$$

where $\overrightarrow{F}_{conn}$ is the force that keeps a UAV connected to its neighbors, $\overrightarrow{F}_{AP}$ is the force that keeps a SN in the range of an AP or a RN, and $\overrightarrow{F}_{CA}$ is the repelling force that keeps the UAV away from obstacles (i.e., other UAVs). The magnitude and direction of each force are determined by the strength of the communication between each pair of UAVs, measured by the RSSI. $\overrightarrow{F}_{conn}$ is calculated as follows:

$$\overrightarrow{F}_{conn} = \sum_{k=1}^{d} \overrightarrow{F}_{conn_k}, \tag{2}$$

$$\overrightarrow{F}_{conn_k} = q|\Delta RSSI|\overrightarrow{d_{0_k}}. \tag{3}$$

For each UAV, the d neighboring UAVs with the best connections are chosen. For each pair, $\overrightarrow{F}_{conn_k}$ is computed. $\Delta RSSI$ measures how far the current RSSI ($RSSI_{curr}$) is from the maximum RSSI ($RSSI_{max}$) or minimum RSSI ($RSSI_{min}$) threshold, while $\overrightarrow{d_{0_k}}$ is the normalized directional vector between the two UAVs, and q determines if the force is attractive or repulsive. If $RSSI_{curr}$ is above $RSSI_{max}$, the force is repulsive, while if it is below $RSSI_{min}$, the force is attractive. The force is zero only when $RSSI_{curr}$ is between $RSSI_{min}$ and $RSSI_{max}$. The resulting force $\overrightarrow{F}_{dir}$ gives the direction in which the UAV has to move in the current time step.

Di Felice et al. [4] propose a distributed algorithm based on VSFs to give coverage to a set of isolated GNs. The mobility of UAVs is subject to three rules: connectivity of the aerial mesh must be preserved, QoS of the links must be guaranteed and the covered GNs must be preserved. These requirements are obtained by moving each UAV according to a virtual force $\overrightarrow{R}$ given by the sum of the spring forces associated with each wireless link established by the UAV. Each force is computed as follows:

$$\overrightarrow{F} = -k \cdot \delta, \tag{4}$$

where k is the stiffness of the spring and δ is its displacement, computed as a function of the link budget $LB(i, j)$ measuring the quality of the wireless link (i, j). For Air-to-Air (AtA) links, k is a fixed parameter. For Air-to-Ground (AtG) links, k is dynamic and defined as follows:

$$k_{AtG} = \frac{n_i}{max(n_j) \forall j \in Neigh_i}. \tag{5}$$

UAVs connecting more GNs oppose more resistance to movements that might reduce the number of covered nodes. At each time step ΔT, the resulting force $\overrightarrow{R}$ is computed for each UAV and then, if the force magnitude is above a certain threshold, the UAV moves with constant speed in the new direction. An exploration phase is also triggered each T_{scout} seconds. All UAVs with no other UAV in their visibility zone become scout nodes with probability p_{SCOUT} and an attractive spring force is computed between them and the center of the least visited cell. In [11], the authors perform a comparison with a centralized optimization scheme, which concludes that the distributed algorithm can achieve a performance that is similar to the centralized one.

Reynaud et al. [12] propose a virtual force based scheme in a scenario of detection and location of *vespa velutina*. As pursuit UAVs move away from the control node to pursue *vespa velutina* individuals, relay UAVs form a communication chain that provide multihop connectivity to pursuit nodes. Within a communication chain, distances between each two neighbors determine if the force between them is attractive, friction or repulsive, in order to keep the chain connected. A development of this scheme is proposed in [13] to provide area coverage in zones where the network infrastructure is damaged or non-existent. The concept of communication chain is now applied between a source and a destination of data. Besides the already mentioned attractive, friction and repulsive forces, relay UAVs are now subject to alignment forces, which seek to straighten the communication chains, making them as coincident as possible to the line segment that connects the source and the destination nodes. This will make the chain more efficient by minimizing the number of nodes. According to the proposed scheme, the UAVs that do not belong to a chain become survey UAVs moving randomly to locate potential source and destination nodes. The authors do not address the problem of topology management under multiple simultaneous source-destination pairs. For example, they do not consider fusing or gluing together closely located or crossed chains in order to minimize the overall number of relays. They also do not explain how relays are assigned to chains when there are multiple possibilities.

3. The LoRaUAV System

The objective of LoRaUAV is to provide coverage to mobile GNs using a fleet of UAVs, whose purpose is to relay the collected data to a BS through an ad hoc network established between its members. Each UAV must be able to adjust its position to reflect the changing topology of the GNs and, at the same time, maintain connectivity with the BS. Disconnections, although difficult to avoid, must be minimized, and recovery measures must be set up to recover the connectivity. LoRaUAV provides mechanisms to tackle all these problems. Some assumptions are made:

1. All UAVs fly at the same altitude in a 2D plane;
2. All UAVs can move with constant speed in any direction;
3. UAVs periodically exchange their position and the list of covered GNs with their neighbors;
4. UAVs have access to the received power of neighboring UAVs and GNs or can estimate it based on their position;
5. The position of the BS is known by all UAVs;
6. The minimum data rate (and hence maximum SF) acceptable for each LoRaWAN link is a configuration parameter.

3.1. Architecture

The architecture of LoRaUAV consists of a two-layer system: the first layer is composed of GNs that transmit data using LoRaWAN, while the second layer is composed of a swarm of relay drones communicating over an WiFi ad hoc network. LoRaUAV is composed of three entities: GNs, UAVs and a BS. GNs are equipped with devices that aggregate data coming from various sensors and transmit it using a LoRaWAN module. The UAVs must be equipped with all the necessary sensors, controls and software for navigation and stabilization, e.g., speed meters and accelerometers. We assume that drones already come with all the necessary components and with a software API to interact with them. The UAVs considered in this work are multi-rotors able to hover over a location or target. The UAVs carry a module that integrates a LoRaWAN Gateway (GW) chip and a WiFi chip supporting ad hoc communications. End-to-end packet delivery over WiFi is accomplished thanks to a traditional TCP/IP stack, with the assistance of a proactive MANET routing protocol. The module also contains a relay application, responsible for adapting LoRaWAN packets received by the GW, and sending them over the ad hoc network to the destination. The most important part of the system is the VSF distributed mobility algorithm, responsible for planning the movements of the drone according to virtual spring forces on the basis of parameters collected from the GPS, routing tables and messages exchanged with GNs and other UAVs. Therefore, the algorithm needs to interface with almost all UAV systems. The BS contains the same WiFi equipment as a UAV, runs the same MANET routing protocol and integrates a LoRaWAN Network Server (NS) to manage the network. Other LoRaWAN entities (Application Servers (ASs) and Join Server (JS)) can be placed anywhere provided that a form of network connectivity is available (e.g., satellite or terrestrial link).

3.2. VSF Mobility Algorithm

LoRaUAV implements a force-based distributed mobility algorithm that can push or pull an UAV closer or further from other UAVs or GNs. In this way, UAVs are kept within communication range of each other, collisions are avoided and the objective of covering the GNs is pursued. The implemented VSF algorithm is an adapted version of the algorithm described by Di Felice et al. in [4]. Modifications affect the weights of AtA and AtG forces. The algorithm runs at fixed regular intervals ΔT. At each time step, each UAV generates a set of virtual springs, having one end attached to the UAV that generated them, and the other end attached to either a GN (AtG spring) or another UAV (AtA spring). For each spring, the force F_{ij} between node i and node j is calculated as follows:

$$\overrightarrow{F_{ij}} = K \cdot (LB_{ij} - LB_{req}) \overrightarrow{x}, \tag{6}$$

where K is a proportionality factor, LB_{ij} is the real or estimated link budget between node i and node j, and LB_{req} is the required link budget. Link budget measurements give a better indication of the quality of the link and allow to better select the parameters according to application-specific QoS requirements. The total force F_{tot}^i applied to UAV i is:

$$\overrightarrow{F_{tot}^i} = \sum_{j=0}^{N} \overrightarrow{F_{ij}}, \tag{7}$$

where N is the number of springs originating from UAV i. AtG springs are established with all the GNs that have sent to the UAV at least one update in the last t_{AtG} seconds. AtA springs, instead, are established with any one-hop neighboring UAV. AtA and AtG springs are only established with direct neighbors, thus lowering the computations and the complexity of the interactions. The K parameter is fundamental to determine the weight and priority of each spring force F_{ij}. For AtA springs, K is updated according to the following expression:

$$K_{AtA} = K_p \left(\frac{N_{neighs}^{max}}{n_{neighs}} \right), \tag{8}$$

where $K_p > 0$ is a scaling factor, N_{neighs}^{max} is the maximum expected number of neighboring UAVs and n_{neighs} is the current number of neighboring UAVs. Equation (8) allows to scale the magnitude of the AtA forces on the basis of the relative number of neighboring UAVs. In this way, UAVs with few neighbors have a higher tendency to stick to the formation, while UAVs with many neighbors exhibit the opposite behavior, therefore assigning more relevance to AtG springs. A good value of N_{neighs}^{max} typically resides in the interval $[6, 8]$, since UAVs tend to arrange themselves in hexagonal grid formations when subject only to AtA forces. If the number of drones is small, N_{neighs}^{max} can be approximated with the total number of UAVs. The K parameter takes a different value at each time step for each AtG spring attached to GN j:

$$K_{AtG}^j = \frac{u_{max}}{u_j}, \tag{9}$$

where u_{max} is the highest number of neighboring UAVs sharing one or more covered GNs and u_j is the current number of UAVs covering GN j. This relation gives more priority to the GNs that are covered by less UAVs, and drones are therefore expected to distribute themselves uniformly between the GNs that are in range.

3.3. Connection Recovery and Maintenance (CRM) Algorithm

In LoRaUAV, partitions of the UAVs are difficult to avoid. Disconnections might be caused by difficult environmental conditions or by one or more UAVs that go too far from nearby UAVs due to strong attractive forces. Some mechanisms are therefore integrated in the system to recover the connectivity with the UAV formation and with the BS. After the connection to the BS is recovered, a connection maintenance procedure is triggered. Figure 2 shows the developed CRM algorithm. The CRM algorithm has three mobility modes: the VSF mobility described in Section 3.2, the Network Recovery Mobility (NRM), which consists in moving the drone towards the BS to recover the connection, and the Stationary Mobility (SM), which forces the drone to hover over a location. The *pause* and *persist* parameters are used to select the mobility mode. The *pause* parameter is a positive integer that specifies for how many time intervals the UAV has to remain stationary. When a connection with the BS is recovered thanks to the NRM, the SM is activated and kept active for *pause* $\times \Delta T$ time intervals in order to avoid further disconnections. Every UAV sets *pause* according to the following expression:

$$pause = \left\lfloor P_{max} \frac{R_{max}}{D_{BS}} \right\rfloor, \tag{10}$$

where P_{max} is the maximum number of pause intervals, R_{max} is the maximum communication range and D_{BS} is the current distance from the BS. R_{max} is updated progressively thanks to the position information periodically exchanged with neighboring UAVs. Equation (10) assigns a more conservative behavior to drones that are closer to the BS and therefore more likely to cause the disconnection of a larger part of the network. On the contrary, peripheral UAVs can act more freely, since their temporary disconnection is less likely to cause the partition of other parts of the network. The *pause* parameter is ignored only when the *load* parameter is zero. This parameter measures the number of relayed packets during a time window of fixed length t_{load} and it can therefore be used to select inactive UAVs free to move without the constraints imposed by the CRM algorithm. The *persist* parameter is a positive integer that specifies for how many time intervals the UAV has to keep using the VSF mobility even without a connection to the BS, with the hope that a connection can be recovered thanks to the rearrangement of other UAVs. The *persist* parameter is set according to the following expression:

$$persist = \left\lfloor (hops - 1) \frac{D_{BS}}{R_{max}} \right\rfloor, \tag{11}$$

where *hops* is the last recorded hop distance between the UAV and the BS. Equation (11) allows peripheral UAVs to be dragged by VSF forces for a longer time, even if they are isolated. On the contrary, UAVs closer to the BS are less tolerant to disconnections and try to recover a connection faster. It is possible that a group of UAVs, sharing some covered GNs, gets isolated from the swarm. In this case, the typical *persist* behavior would not exploit the redundant UAVs to recover the connectivity. Instead, all the UAVs would trigger the NRM at almost the same time and all would loose the connection with the covered GNs. To avoid this situation, UAVs that are covering a set of GNs determine, after loosing the connection, the neighboring UAV with which they share the highest percentage of GNs. If this percentage is above a certain threshold p_{shared}, one of the two UAVs is redundant. Every UAV has a fixed ID that has been assigned before the operation. The UAV with the lowest ID among the two cuts by half its *persist* intervals, while the other UAV doubles those intervals. In the next time steps, the UAV with the lowest value of *persist* triggers the NRM before the others and repositions itself closer to the BS. This helps to recover a connection with the UAV swarm without loosing the GNs.

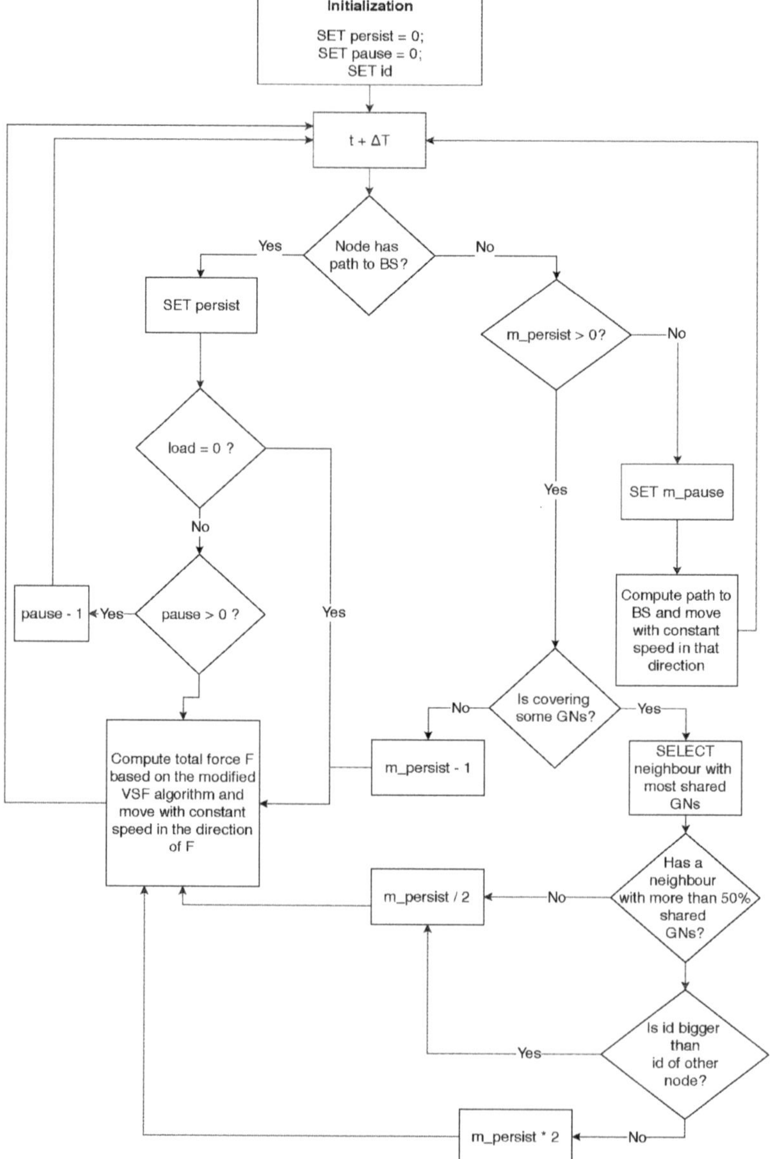

Figure 2. CRM algorithm.

3.4. Movement Prediction (MP) Algorithm

The MP algorithm is a solution to recover isolated GNs based on a combination of movement prediction and virtual forces. This last characteristic allows to easily integrate the solution in the system and to preserve the simplicity of the LoRaUAV distributed approach. The algorithm is based on entities called *holograms*, whose purpose is to signal to UAVs the estimated position of isolated GNs.

A hologram contains a pair of spatial coordinates that can be used to compute a virtual AtG force between the hologram itself and other UAVs. Since GNs are moving, the information about their last known position is not enough to create useful holograms. However, if GNs keep a regular direction and speed for some time, the future position of GNs can be predicted from previous recorded positions and the hologram location can be updated accordingly at each time step. In LoRaUAV, the prediction is based on simple kinematic rules. In order to estimate the kinematic rule parameters, the algorithm uses the GN position updates. The time must thus be treated as a sequence of discrete time steps. If we consider a GN i at time t, its future coordinate $x_{i(t+1)}$ is:

$$x_{i(t+1)} = x_{it} + v_{it}^x + \frac{1}{2}\hat{a}_{it}^x \cdot \Delta t, \tag{12}$$

where v_{it} and $\hat{a}_{it}^x$ are respectively the velocity at time t and the expected variation of velocity (acceleration) within the update time interval Δt. The latter also corresponds to the interval between t and $t+1$. As will be seen in Equation (13), v_{it} is computed as the difference between x_{it} and $x_{i(t-1)}$. As such, it already corresponds to the displacement within Δt. The computation of all the necessary variables is based on the finite difference method and only requires the knowledge of the three most recent positions of a node, so that:

$$\begin{cases} v_{it}^x = x_{it} - x_{i(t-1)} \\ v_{i(t-1)}^x = x_{i(t-1)} - x_{i(t-2)} \\ v_{i(t-2)}^x = x_{i(t-2)} - x_{i(t-3)}, \end{cases} \tag{13}$$

$$\begin{cases} \hat{a}_{it}^x = a_{i(t-1)}^x + \Delta a_{i(t-1)}^x \cdot \Delta t \\ \Delta a_{i(t-1)}^x = \dfrac{a_{i(t-1)}^x - a_{i(t-2)}^x}{\Delta t} \\ a_{i(t-2)}^x = \dfrac{v_{i(t-1)}^x - v_{i(t-2)}^x}{\Delta t} \\ a_{i(t-1)}^x = \dfrac{v_{it}^x - v_{i(t-1)}^x}{\Delta t}, \end{cases} \tag{14}$$

It should be noted that, as expected, velocity and acceleration are still related as follows:

$$v_{i(t+1)}^x = v_{it}^x + \hat{a}_{it}^x \cdot \Delta t, \tag{15}$$

A similar reasoning applies to coordinate y_i. In the current version, in order to simplify the problem, only GNs with constant velocity are considered, so that only the two latest positions, separated at most by t_r seconds, are required. UAVs are responsible for keeping track of lost GNs (in the last t_{lost} seconds) and for sending to other UAVs the information needed to predict the location of holograms: the last known position of a GN, its velocity vector and the time of its latest recorded position. This information is included in a message called *token*, sent in unicast to every UAV in the routing table. A hologram is created for each token and an expiration timer t_{exp} is attached to it. To avoid disruptions of AtA and AtG forces, only UAVs that are inactive (*load* $= 0$) are influenced by hologram forces. To reduce the amount of exchanged *tokens*, lost GNs can be first clustered in groups using a clustering algorithm. In LoRaUAV, *k-means* [14] is used for clustering and *k-means++* [15] is used to initialize the centroids and speed up the convergence. The *k-means* algorithm is run multiple times with different values of k, with k ranging from 2 to k_{max}, and the quality of the clustering is then assessed through the *average silhouette index*, a metric that measures the similarity among members of the same cluster and their dissimilarity from members of other clusters. The silhouette $s(i)$ of the data point i is calculated as follows:

$$s(i) = \frac{b(i) - a(i)}{\max\{a(i), bi\}}, \tag{16}$$

where $a(i)$ is the average distance of i from the members of its own cluster and $b(i)$ is the lowest average distance of i from the members of neighboring clusters. The index gets values in the interval $[-1, 1]$. Values closer to 1 indicates that the data point i belongs to a good cluster. Data points belonging to a cluster with only one point have a silhouette value of zero. The average s_{avg} of all $s(i)$ is compared to the minimum threshold s_{thr} to assess the overall quality of the clustering. If no good clustering is found, all lost GNs are assigned to a single cluster. A *token* is created for each cluster and sent to the other UAVs. The token contains the center of mass, the average velocity and the time of the most recent update of each cluster.

4. Simulation Model

The LoRaUAV system was developed and evaluated using the ns-3 [16] network simulator. ns-3 has been chosen for its modular architecture, its speed and for its support of all the protocols that are part of LoRaUAV, including modules for simulating LoRaWAN networks. Of the available LoRaWAN modules for ns-3, the one developed by Magrin et al. [17] has been chosen. The main modeling choices are described in this section. The LoRaUAV model developed for this paper is available at the following link: https://github.com/marcostellin/ns3lorauav.

4.1. Channel Propagation Models

AtA links are modeled with a Friis propagation model. It is assumed that the monitored area is flat and no obstacles are present in the LoS of the UAVs at any moment. Since UAVs operate several meters above the ground, the LoS assumption is acceptable. Under these conditions, the Friis propagation model represents a reasonable choice [18]. The path loss PL in dB is therefore computed as follows:

$$PL = G_t + G_r - 10 \cdot \log\left(\frac{(4\pi d)^2 L}{\lambda^2}\right), \tag{17}$$

where G_t and G_r are antenna gains in dBi, λ is the wavelength in meters, d is the distance between the transmitter and the receiver in meters and L is the system loss. G_t and G_r are set to 0 (omnidirectional antennas) and L is set to 1.

In the considered scenario, GNs move in an environment rich of vegetation. Under these conditions, the LoS between GNs and UAVs is not guaranteed at all times, and foliage and trunks produce considerable absorption, scattering and diffraction. In this work, the AtG channel is modeled with a Log Distance path loss model described by the following equation:

$$PL = L_0 + 10 \cdot n \cdot log_{10}\left(\frac{d}{d_0}\right), \tag{18}$$

where n is the path loss exponent, d is the distance in meters from the receiver, and d_0 and L_0 are respectively the reference distance and the reference loss for d_0. These parameters are estimated from the LoRaWAN RSSI measurements obtained in a forest environment and collected by Iova et al. in [19] ($L_0 = 32.22$ dbm, $n = 5.2$, $d_0 = 1$ m). The path loss exponent value abstracts the range limitations imposed by fading and shadowing effects. The authors show that a significant range drop is expected when LoRaWAN is used in environments with thick vegetation, even if high SFs are used. The implemented model gives a worst-case estimation of the LoRaWAN expected range that agrees with the real experimental measurements in [19].

4.2. Firefighters Mobility Model

The VSF algorithm adapts the topology of the UAV swarm according to the movements of GNs. To validate the performance of the developed algorithm, it is therefore necessary to assign a suitable mobility model to the GNs. Given the considered application, GNs must act similarly to firefighters during a wildfire operation. To the best of our knowledge, no mobility model has ever been created for this situation and therefore a new one has been developed based on information collected from [20] and [21]. An approximation of this behavior has been implemented. According to this model, each GN i is assigned to a team j, with $j = 0, 1, 2.., N$. Teams can be placed anywhere in the simulation area, which is assumed to be an $S \times S$ square. The whole area is divided vertically in N columns of equal width, one for each team, and each column is in turn split horizontally in R_j equal width rows, with R_j given by:

$$R_j = \lfloor S/d_{rj} \rfloor, \tag{19}$$

where d_r^j is the retreat distance of team j. At this point, every team has an assigned vertical stripe with R_j cells. At $t = 0$, each team selects a random point in the furthest cell from the origin of the scenario in its assigned stripe and moves towards it with constant speed until the destination is reached. At this point, the team members start moving independently and randomly for a certain amount of time t_{rw} (deterministic or random) in a rectangular $s_j^x \times s_j^y$ operation area. When time is over, teams select a new point in the second furthest cell and repeat the previously described process. Teams stop moving when the simulation is over or when they have visited all the cells in their column. This model tries to mimic the progress of a wildfire and the consequent retreat of firefighters to fallback areas where new attacks to the fire are attempted. A representation of the model is shown in Figure 3.

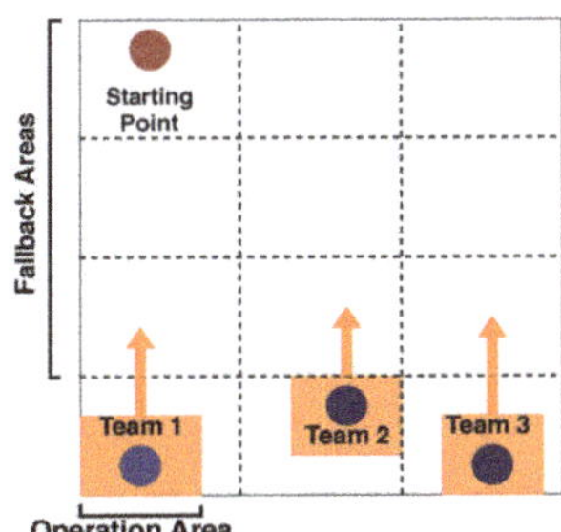

Figure 3. Firefighter team mobility model. Teams start from a starting point and go down to the furthest cell in their assigned column.

5. Simulation Results

An incremental evaluation of the algorithms has been made. All fixed parameters are reported in Table 1. As the objective is to evaluate the performance of the UAV topology control algorithms, the spreading factor of LoRaWAN was fixed to SF7, the worst case, resulting in the shortest AtG communication range. The routing algorithm used in the UAV mesh is Optimized Link State Routing Protocol (OLSR) [22]. The main evaluation metrics are the Average End-to-End Packet Reception Ratio (AE-PRR) and the Average Total Delay (ATD). The first metric measures the average PRR of unique packets at the BS, while the second metric measures the average end-to-end delay, including the buffer delay of uplink packets. Averages consider 100 runs of the same scenario (seed in $[0, 99]$). The 95% confidence interval is also computed assuming a Student's t-distribution of the population. The default simulation parameters are reported in Table 2.

A higher value of the K_p parameter favors the compactness of the aerial mesh, reducing the disconnections. However, it reduces the coverage. K_p should therefore be optimized in each scenario to achieve the best compromise between the desired coverage and delay. This optimization is relegated to future work. In the presented simulation results, K_p is treated as an independent variable.

Table 1. List of fixed simulation parameters.

LoRaWAN		WiFi and Routing	
GN Tx Power	14 dBm	PHY protocol	IEEE 802.11g
Class	Class A	Modulation	ERP-OFDM
Bandwidth	125 kHz	Frequency	2.4 GHz
Frequency	868.1 MHz	Tx Power	16.02 dBm
SF	7	Rx Sens.	-99 dBm
GW Sens. (dBm)	-124 (SF 7)	Bitrate	12 Mbps
Packet size	10 bytes	MANET protocol	OLSR
Packet period	30 s		
VSF Algorithm		**CRM Algorithm**	
ΔT	10 s	P_{max}	30
t_{AtG}	40 s	p_{shared}	0.5
LB_{req}	20 dbm	t_{load}	3 min
MP Algorithm		**Firefighter Mobility Model**	
t_{lost}	3 min	$s_j^x \; \forall j$	300 m
t_r	40 s	$s_j^y \; \forall j$	100 m
t_{exp}	10 min	d_r^j	$\mathcal{U}(50, 250)$ m
k_{max}	3	t_{rw}	5 min
s_{thr}	0.6	BS position	$(S/2, 300)$

Table 2. Default simulation parameters.

N. of UAVs	$4, 8, 12$
N. of teams	1-5
GNs per team	20
Sim. Time	2000 s
Area side	2000 m
K_p	$[1, 50]$

5.1. Comparison of LoRaUAV VSF and DF VSF Algorithms

A comparison between the LoRaUAV VSF algorithm (see Section 3.2) without extensions, and the VSF algorithm described by Di Felice et al. [4] (DF algorithm) has been performed. Before starting the analysis, it must be clarified that the K_p parameter coincides with the proportionality factor of K_{AtA} in the LoRaUAV VSF algorithm and it coincides, instead, with K_{AtA} itself in the DF VSF algorithm. An AE-PRR comparison between the two algorithms is shown in Figure 4 for a scenario with three and five teams of GNs. For clarity, the confidence interval is not shown. It is possible to notice that the AE-PRR of both algorithms follows approximately the same trend for all configurations, presenting the same regions of increase and decrease and the same range of optimal values. However, the AE-PRR produced by the LoRaUAV VSF algorithm seems slightly better than the one generated by the DF VSF algorithm for most K_p values. To verify this hypothesis, the curves generated by the two algorithms have been compared by computing the point by point euclidean distance between them for each configuration. On average, LoRaUAV produces an AE-PRR increment which lies between 0.03% and 4.6% independently of the value of K_p that is used. The developed VSF algorithm produces more benefits when a large number of teams is deployed (e.g., five teams). However, for most K_p values, improvements fall inside the confidence interval of the DF VSF algorithm. It is therefore not possible to assert that the LoRaUAV VSF algorithm performs better in general. In fact, changing just the weights of the forces is not by itself sufficient to achieve a significant AE-PRR increase. Nevertheless, even if small and limited to particular configurations, the improvements introduced by the developed algorithm come at no significant additional cost, apart from a delay penalty affecting some packets and correlated with the higher PRR. As such, at this stage, basic LoRaUAV becomes at least validated in comparison with DF. Further comparison regarding the frequency and duration of disconnection times is provided in Section 5.4.

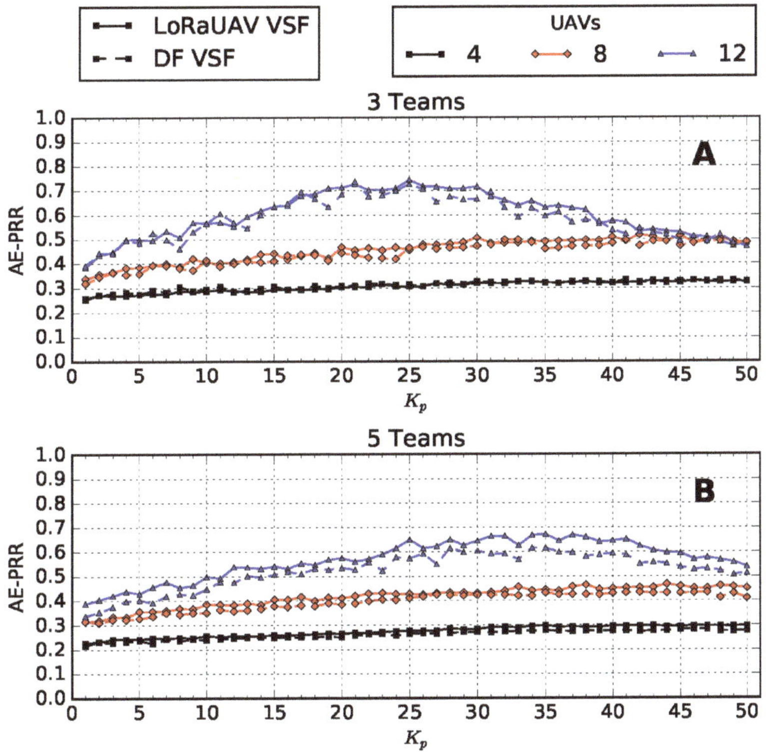

Figure 4. Average End-to-End Packet Reception Ratio (AE-PRR) comparison when the LoRaUAV Virtual Spring Force (VSF) algorithm and the DF VSF algorithm are used in a scenario with three (**A**) and five (**B**) teams of 20 GNs.

5.2. Study of the Impact of the CRM Algorithm

The CRM algorithm has been developed to increment the coverage provided by the UAV swarm, without adding more UAVs. In fact, the AE-PRR obtained with just the VSF algorithm is affected by a high percentage of lost packets. Depending on the application-specific QoS requirements, this might be a major drawback. It is shown that the CRM algorithm effectively improves the AE-PRR performance of the system in comparison with the basic LoRaUAV VSF algorithm operating alone. In the first simulations, the size of the monitored area (2500 m of area side and 3000 s of simulation time) has been chosen specifically to favor CRM and hence to show its intent. A visual comparison of a sample scenario with three teams of GNs is plotted in Figure 5. In all the performed simulations, the CRM algorithm significantly improves the AE-PRR, with most of the increments lying between 14% and 26% (2–5 teams), but with peaks of more than 50% (one team). In absolute terms, the achieved AE-PRR is low. The ATD seems to be the QoS metric that is most negatively affected by the CRM algorithm, as shown in Figure 5B. However, a deeper analysis shows that the ATD metric is greatly influenced by a small number of packets with large delays. In fact, the CRM algorithm allows for more and longer disconnections in order to achieve a higher PRR. Some packets may therefore stay in the buffer for a long time before being sent. To prove this hypothesis, 10 individual simulations have been run for a scenario having 12 UAVs, three teams of GNs and $K_p = 25$. The total individual delays of all received packets have been recorded. The percentage of buffered packets more than doubles when the CRM algorithm is active (from ≈2–3% to ≈4–8%) and their permanence in the buffer increases drastically (from ≈15–34 s to ≈34–181 s). If buffered packets are not considered, the average experienced delay is ≈55 ms, a value that is far lower than the reported ATD values.

The second set of simulations has considered the more reasonable default settings in Table 2. The results are very encouraging, as concluded from the comparison between Figure 6A,B, where CRM significantly improves the performance of VSF, usually by more than 30%.

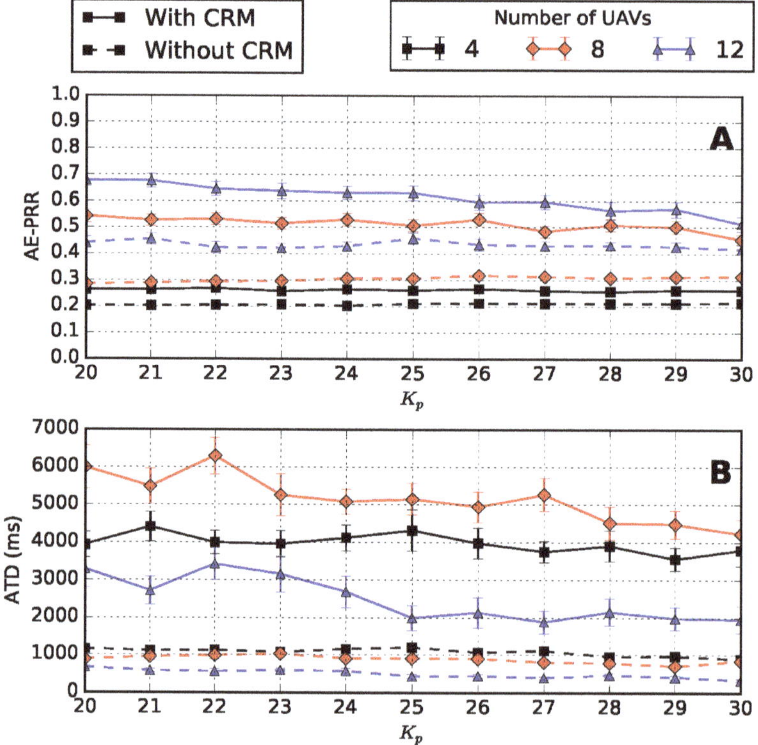

Figure 5. AE-PRR (**A**) and Average Total Delay (ATD) (**B**) comparison when the CRM algorithm is activated in a scenario with three teams of 20 GNs.

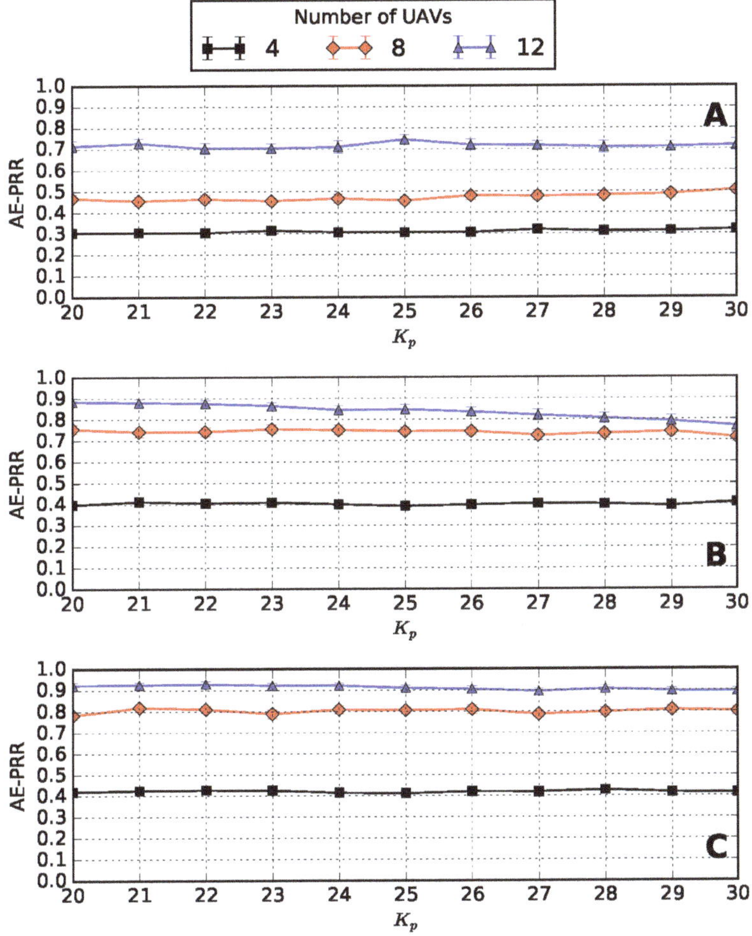

Figure 6. AE-PRR with VSF (**A**), VSF+CRM (**B**) and VSF+CRM+MP (**C**) (no team splits) in scenario with the same parameters of Section 5.1.

5.3. Study of the Impact of the MP Algorithm

The MP algorithm represents the last attempt of the system to recover a connection between the aerial mesh and the GNs that went out of range. For this reason, the PRR is used as the main evaluation metric. In order to increase the triggering chances of the MP algorithm, the firefighters mobility has been extended to support the splitting of teams. With this modification, a team can be configured to split in half during the simulation, so that an UAV that is covering alone a team of GNs is forced to create at least one hologram when half of the team eventually goes out of range. Given this modification, it is of particular interest to observe the behavior of the MP algorithm in two situations: a more conventional setting, where team splitting is not active, and in a more challenging setting where teams can split. The end-to-end PRR is collected individually for all the 100 runs of the same scenario and a direct comparison between corresponding runs of different scenarios is performed. The results of the comparison are reported in Table 3, considering three teams and 2500 m of area side and 3000 s of simulation time. The first observation is that the MP algorithm is not always beneficial to the system. In fact, a relevant number of runs shows a PRR that is, on average, even 7% lower than the one achieved without the MP algorithm. The loss is more marked in scenarios in which teams do not split. This is caused by the fact that the strategy of recovering the connectivity with lost GNs is not always beneficial to the UAV formation since it does not follow any global optimality criteria. However, the simultaneous activation of the MP and CRM algorithms is beneficial in most of the situations. In fact, the average PRR undergoes a considerable increment in at least 66 % of the runs when teams do not split and 71 % of the runs when teams split. In some cases, the gain that is obtained is more than double of the loss observed in analogous scenarios without MP algorithm. When reverting to the default simulation parameters, the activation of the MP algorithm and of the CRM algorithm produces the results shown in Figure 6C. Even if small, the improvements allow to achieve AE-PRR levels above 0.9 with 12 UAVs and between 0.8 and 0.9 with 8 or 10 UAVs. Similar tests have been run for the VSF+MP configuration with no noticeable advantages.

Table 3. Comparison between runs with VSF + CRM + MP and runs with only VSF + CRM.

No Split				
UAVs	**Runs**		**Avg MP PRR Gain/loss**	
	Better w/ MP	Better w/o MP	Better w/ MP	Better w/o MP
4	58%	42%	+2.1%	−1.5%
6	53%	47%	+9.0%	−7.1%
8	66%	34%	+13%	−6.6%
10	77%	23%	+15%	−7.0%
12	86%	14%	+18%	−4.0%
Split				
UAVs	**Runs**		**Avg MP PRR Gain/loss**	
	Better w/ MP	Better w/o MP	Better w/ MP	Better w/o MP
4	78%	22%	+4.3%	−2.4%
6	71%	29%	+8.9%	−5.9%
8	69%	31%	+11%	−5.8%
10	78%	22%	+11%	−5.2%
12	91%	9%	+15%	−1.9%

5.4. Frequency and Duration of Disconnections

LoRaUAV is subject to the inevitable disconnection of some GNs. In this section, the frequency and duration of such disconnections is investigated and analyzed. The simulation parameters, which are reported in Table 4, were configured so that they fall within the default parameter ranges.

Table 4. Parameters used for testing the frequency and duration of Disconnections.

Parameter	Values
Number of UAVs	12
Number of teams	3
GNs per team	20
K_p	25
Total simulation time	3000 s
Simulation area	2000×2000 m

The simulation script has been adapted to periodically check the GNs that are covered by each UAV. This information is then used to compute, for each GN, the number and duration of disconnection periods. The results of 100 runs are aggregated and reported in this section. The total average disconnection time, computed as the average of the sums of all disconnection periods of all the performed runs, is reported in Table 5 for all the algorithms. It can be noted that, on average, the activation of the CRM and of the MP algorithms effectively reduces almost by half the total experienced disconnection time in comparison with both VSF algorithms. The VSF + CRM and the VSF + CRM + MP configurations perform almost identically. However, if we separate the total average disconnection times of the different teams, it can be seen that, while in the configuration with just the CRM, the average disconnection time is split almost equally among teams ($\approx$4000 s), the configuration with the MP algorithm favors team 0 and team 1 ($\approx$2200 s) at the expense of team 2 ($\approx$7500 s). A visual inspection of the runs where long disconnection periods are experienced showed that team 2 tends toward a region of the map that is more difficult to access for the aerial mesh. Disconnections of the other teams are solved earlier in the simulation, thanks to the MP algorithm. This leads the UAV mesh to steer towards the location of team 0 and team 1, thus making recovery of team 2 less likely. As already stated, the current version of the MP algorithm suffers from the lack of global optimality criteria that can better direct the connection recovery efforts.

The histogram of Figure 7 details the results for different combinations of algorithms. In order to better visualize the results, disconnection periods are grouped in bins of different lengths and colors are used to distinguish between different teams. The first observation that can be made is that all the algorithms are subject to partitions, so that GNs might get isolated even for long periods of time. This is particularly evident when the VSF algorithms are used alone. In fact, for both DF VSF and LoRaUAV VSF, the number of disconnections above 500 s constitutes a big share of the total experienced disconnections. In this regard, the DF VSF algorithm seems to experience less disconnections than the LoRaUAV VSF algorithm. This can be explained by the reduced coverage offered by the DF algorithm and the consequent presence of fewer but longer disconnection periods. In fact, a more in depth analysis of the disconnection periods larger than 500 s of the DF VSF algorithm reveals that several GNs remain isolated for more than 2500 s, while, with the LoRaUAV VSF, isolation periods are no longer than 2000 s, meaning that the mesh is able to cover the GNs for a longer time before loosing them definitely. Nonetheless, both VSF algorithms show low performances and a high chance of almost permanent isolation of GNs. This claim is supported by the low AE-PRR results obtained by the VSF algorithms in all the previous sections of this chapter. The CRM algorithm improves the situation. One first observation is that the algorithm is fairer towards teams. In fact, disconnection periods of the same duration are approximately uniformly distributed among teams. This is caused by a better average distribution of the UAV mesh. The second observation is that disconnections are more frequent if compared with the LoRaUAV VSF algorithm operating alone. However, disconnections are shorter and the number of disconnection periods above 500 s is reduced by $\approx$1000 units overall, with team 0 and team 1 experiencing the most marked improvements. This means that, albeit partitions are still possible, they either happen much later in the simulation or they are recovered sooner by the system. The introduction of the MP algorithm on top of the CRM algorithm gives some interesting results. First, the frequency of shorter disconnection periods increases in comparison with the results obtained with just the CRM algorithm, especially in the interval $[0, 30]$ seconds. This increase can be explained by the numerous connection recovery attempts, some of them successful, others only partially successful because the connection is lost again after some time. The disconnection periods above 500 s remains approximately the same obtained with the CRM algorithm, but with a huge disproportion towards team 2 members for the reasons already explained before in this section.

Table 5. Total average disconnection time and relative confidence interval obtained with different algorithm combinations.

Algorithm	Total Average Disconnection Time (s)
DF VSF	23,199 $\pm$ 3350
LoRaUAV VSF	21,826 $\pm$ 2976
LoRaUAV VSF + CRM	12,365 $\pm$ 2258
LoRaUAV VSF + CRM + MP	11,876 $\pm$ 1600

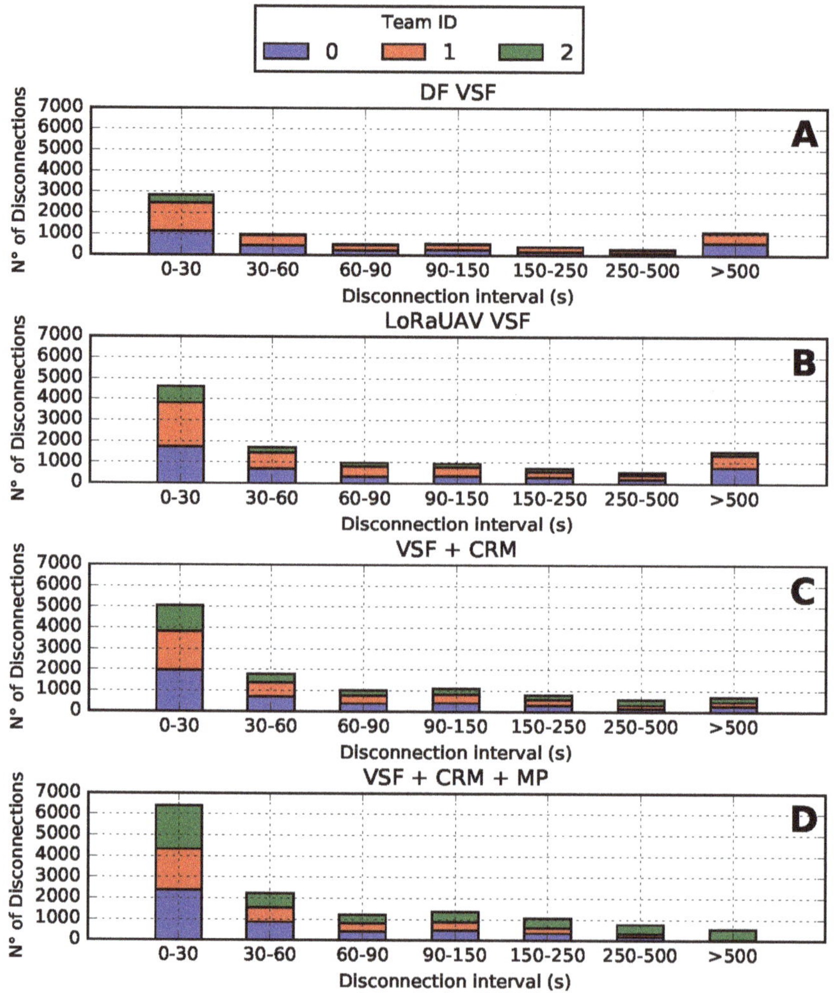

Figure 7. Frequency and duration of disconnections for DF VSF (**A**), LoRaUAV VSF (**B**), LoRaUAV VSF+CRM (**C**) and LoRaUAV VSF+CRM+MP (**D**).

6. Conclusions

In this paper, a new system called LoRaUAV has been proposed. The objective of the system is to extend the coverage offered by a BS to a set of mobile GNs through UAV relay GWs. This is achieved thanks to a network that consists of a LoRaWAN segment for GNs-GW communications and a WiFi ad hoc network for GWs-BS communications. The core of the system is the distributed LoRaUAV mobility algorithm based on VSFs, designed to adapt the UAV swarm to the changing topology of GNs. The basic VSF mobility is expanded by the CRM and MP algorithms, designed, respectively, to increase the GN coverage through a better UAV distribution and to restore the connection of isolated GNs through movement prediction. The developed algorithms were evaluated through a ns-3 custom-built *loravsf* module. Simulations were run on a model of a wildfire disaster scenario. The performance of each algorithm has been assessed through relevant QoS metrics: the average end-to-end PRR (AE-PRR) and the average delay (ATD). The LoRaUAV VSF algorithm has also been compared with another similar VSF algorithm proposed in the literature. It has been shown that basic LoRaUAV VSF achieves a small average AE-PRR improvement that affects only some values of K_p, with more significant gains when 4 or 5 teams are deployed. The most significant AE-PRR improvements are obtained thanks to the CRM algorithm. In fact, the AE-PRR shows a double-digit increment in most of the tested configurations. The ATD is the metric that is more negatively affected, but long delays only affect a small percentage of packets. The synergy between the CRM and MP algorithms produces another AE-PRR improvement, particularly significant when teams split. The developed extensions to the basic VSF algorithm were also demonstrated to effectively help in reducing almost by half the isolation time of GNs.

Future evolutions of this work will consider the impact of LoRaUAV dynamics on the Adaptive Data Rate mechanism of LoRaWAN, dynamic K_p optimization, more realistic UAV models, more sophisticated prediction algorithms (e.g., Kalman-based tracking), integration with centralized meta-heuristic optimization algorithms (e.g., [6]), downlink messages and the existence of multiple BSs.

Author Contributions: Conceptualization, M.S., A.G.; methodology, M.S., S.S., A.G.; software, M.S.; validation, M.S., S.S., A.G.; formal analysis, M.S., A.G.; investigation, M.S.; resources, A.G.; data curation, S.S., A.G.; writing–original draft preparation, M.S., S.S., A.G.; writing–review and editing, M.S., S.S., A.G.; visualization, M.S., S.S., A.G.; supervision, S.S., A.G.; project administration, A.G.; funding acquisition, A.G. All authors have read and agreed to the published version of the manuscript.

Funding: This work has been supported by Portuguese national funds through Fundação para a Ciência e Tecnologia (FCT) with reference UIDB/50021/2020, by Fundação Calouste Gulbenkian, and also by FITEC-Programa Interface, with reference CIT "INOV-INESC Inovação-Financiamento Base".

Conflicts of Interest: The authors declare no conflict of interest.

References

1. Raza, U.; Kulkarni, P.; Sooriyabandara, M. Low Power Wide Area Networks: An Overview. *IEEE Commun. Surv. Tutor.* **2017**, *19*, 855–873. [CrossRef]
2. LoRa Modulation Basics. Available online: https://www.semtech.com/uploads/documents/an1200.22.png (accessed on 3 June 2018).
3. LoRaWAN 1.1 Specification. Available online: https://lora-alliance.org/sites/default/files/2018-04/loraw antm_specification_-v1.1.png (accessed on 3 June 2018).
4. Felice, M.D.; Trotta, A.; Bedogni, L.; Chowdhury, K.R.; Bononi, L. Self-organizing aerial mesh networks for emergency communication. In Proceedings of the 2014 IEEE 25th Annual International Symposium on Personal, Indoor, and Mobile Radio Communication (PIMRC), Washington, DC, USA, 2–5 September 2014; pp. 1631–1636.
5. Caillouet, C.; Razafindralambo, T. Efficient deployment of connected unmanned aerial vehicles for optimal target coverage. In Proceedings of the 2017 Global Information Infrastructure and Networking Symposium (GIIS), St. Pierre, France, 25–27 October 2017; pp. 1–8. [CrossRef]
6. Sabino, S.; Horta, N.; Grilo, A. Centralized Unmanned Aerial Vehicle Mesh Network Placement Scheme: A Multi-Objective Evolutionary Algorithm Approach. *Sensors* **2018**, *18*, 4387. [CrossRef] [PubMed]
7. Kim, D.; Lee, J. Integrated Topology Management in Flying Ad Hoc Networks: Topology Construction and Adjustment. *IEEE Access* **2018**, *6*, 61196–61211. [CrossRef]
8. Almeida, E.N.; Campos, R.; Ricardo, M. Traffic-aware multi-tier flying network: Network planning for throughput improvement. In Proceedings of the 2018 IEEE Wireless Communications and Networking Conference (WCNC), Barcelona, Spain, 15–18 April 2018; pp. 1–6. [CrossRef]
9. Basu, P.; Redi, J.; Shurbanov, V. Coordinated flocking of UAVs for improved connectivity of mobile ground nodes. In Proceedings of the IEEE Military Communications Conference (MILCOM), Monterey, CA, USA, 31 October–3 November 2004; pp. 1628–1634.
10. Goddemeier, N.; Daniel, K.; Wietfeld, C. Role-Based Connectivity Management with Realistic Air-to-Ground Channels for Cooperative UAVs. *IEEE J. Sel. Areas Commun.* **2012**, *30*, 951–963. [CrossRef]
11. Trotta, A.; Felice, M.D.; Bedogni, L.; Bononi, L.; Panzieri, F. Connectivity recovery in post-disaster scenarios through Cognitive Radio swarms. *Comput. Netw.* **2015**, *91*, 68–89. [CrossRef]

12. Reynaud, L.; Guérin-Lassous, I. Design of a force-based controlled mobility on aerial vehicles for pest management. *Ad Hoc Netw.* **2016**, *53*, 41–52. [CrossRef]

13. Reynaud, L.; Guérin-Lassous, I. Improving the Performance of Challenged Networks with Controlled Mobility. *Mob. Netw. Appl.* **2018**, *23*, 1270–1279. [CrossRef]

14. Hartigan, J.A.; Wong, M.A. A k-means clustering algorithm. *JSTOR Appl. Stat.* **1979**, *28*, 100–108. [CrossRef]

15. Arthur, D.; Vassilvitskii, S. K-means++: The Advantages of Careful Seeding. In Proceedings of the Eighteenth Annual ACM-SIAM Symposium on Discrete Algorithms, Society for Industrial and Applied Mathematics, Philadelphia, PA, USA, 7–9 January 2007; pp. 1027–1035.

16. ns-3. Available online: https://www.nsnam.org/ (accessed on 21 September 2018).

17. Magrin, D.; Centenaro, M.; Vangelista, L. Performance evaluation of LoRa networks in a smart city scenario. In Proceedings of the 2017 IEEE International Conference on Communications (ICC), Paris, France, 21–25 May 2017; pp. 1–7. [CrossRef]

18. Khuwaja, A.; Chen, Y.; Zhao, N.; Alouini, M.S.; Dobbins, P. A Survey of Channel Modeling for UAV Communications. *IEEE Commun. Surv. Tutor.* **2018**, *20*, 2804–2821. [CrossRef]

19. Iova, O.; Murphy, A.; Picco, G.; Ghiro, L.; Molteni, D.; Ossi, F.; Cagnacci, F. LoRa from the City to the Mountains: Exploration of Hardware and Environmental Factors. In Proceedings of the 2017 International Conference on Embedded Wireless Systems and Networks, Uppsala, Sweden, 20–22 February 2017; pp. 317–322.

20. National Wildfire Coordinating Group. *Wildland Fire Suppression Tactics Reference Guide*; National Interagency Fire Center: Boise, ID, USA, 1996.

21. Tymstra, C.; Flannigan, M. Living with Wildland Fire: What we Learned from the 2016 Horse River Wildfire. Available online: https://www.frames.gov/files/9615/1190/3222/2017_November_21_Alaska_Fire_Science_Consortium_webinar_Tymstra_Flannigan.png. (accessed on 11 July 2018).

22. Jacquet, P.; Muhlethaler, P.; Clausen, T.; Laouiti, A.; Qayyum, A.; Viennot, L. Optimized link state routing protocol for ad hoc networks. In Proceedings of the IEEE International Multi Topic Conference, Technology for the 21st Century, Lahore, Pakistan, 30 December 2001; pp. 62–68.

MDPI

St. Alban-Anlage 66

4052 Basel

Switzerland

Tel. +41 61 683 77 34

Fax +41 61 302 89 18

www.mdpi.com

Electronics Editorial Office

E-mail: electronics@mdpi.com

www.mdpi.com/journal/electronics